PRAISE FOR *CENTERS OF PROGRESS*

"Chelsea Follett has written a book telling a history of human ingenuity covering 12,000 years! We learn this astonishing epic by reading about the life—and sometimes death—of 40 cities on five continents. There is a common thread linking these urban biographies: people of different cultures meeting in cities doted with a degree of intellectual freedom develop an ability to solve technical and social problems. And because cities communicate, their inventions slowly add up and soon spread across continents, benefiting humanity. *Centers of Progress* provides an optimistic, well-documented view of the world. We badly need this long-term perspective!"

—Alain Bertaud, author of *Order without Design:*
How Markets Shape Cities

"Cities have long been places that foster innovation and flourishing. They could just as well be called 'labor markets,' 'population clusters,' or 'agglomerations' where people merge and, operating from market signals and personal need, produce stuff others find beneficial. Chelsea Follett's book *Centers of Progress* describes the historical role that cities played in such advancement. From medical innovations that sprang off the Nile River in the third millennium BC in Egypt to the defeat of ruinous '-isms' and rise of the 20th-century liberal order, urban centers made it happen. Underlying Follett's work is the case for fostering the market economies that will help cities continue this role in the future."

—Scott Beyer, author of *Market Urbanism:*
A Vision for Free-Market Cities

"Chelsea Follett's *Centers of Progress* makes urban history interesting again, indeed fascinating. These 40 tales help to explain where modern life comes from and give a broader intellectual and historical tour of the world."

—Tyler Cowen, founder of
Marginal Revolution

"The best way to understand progress is to study history: the case studies of how it actually happened. Here are dozens of such stories—concise and readable—from all over the world, which is a reminder that progress can come from anywhere."

—Jason Crawford, founder of
the Roots of Progress

"Chelsea Follett guides us through 40 cities where innovations enabled humanity to progress to longer and better lives. We revisit beloved cities that are well known to us from travel or reading, such as Athens, London, and New York. But we also learn of the achievements of exotic and underappreciated cities, such as Budj Bim, Nan Madol, and Chang'an. Many of Follett's cities were bastions of freedom in a landscape that was otherwise much less free. She reminds us why our beleaguered cities are worth saving."

—Art Diamond, author of
Openness to Creative Destruction:
Sustaining Innovative Dynamism

"By looking at cities as the centers of progress, this book recasts world history as a developing set of ideas and technologies. This is a much welcome alternative to the approaches to the past that see nations, social classes, or institutions as the drivers and carriers of history. Follett reminds us that history is made by people and that when people congregate in cities, they become more innovative and contribute more to historical development. It is a positive story that everyone can learn from."

—Michael J. Douma, co-editor of
What Is Classical Liberal History?

"Cities are our greatest invention. From ancient cities like Jericho and Uruk to Athens and Rome and onward to Vienna, Kyoto, Florence, Berlin, Paris, London, New York, San Francisco, and many more, Chelsea Follett shows how cities power human progress in technology, arts, and across the board. An absolute must read for mayors, urban leaders, businesspeople, and anyone concerned about the future of our cities, economy, and society."

—Richard Florida, author of
The Rise of the Creative Class

"Endlessly fascinating, wide-ranging, and provocative, *Centers of Progress* takes us on a tour of the most creative moments of human history. Chelsea Follett moves us around the globe and across millennia. From the invention of agriculture to the digital revolution, Follett shows us the many ways in which cities have freed the imagination and brought forth new ideas that improved our lives. An inspiring rebuttal to stories of decline, Follett demonstrates that whenever people were free to gather, interact, and innovate, progress followed."

—Jack A. Goldstone, author of
*Why Europe? The Rise of the West in
World History, 1500–1850*

"We shouldn't just study the past to avoid repeating mistakes; we should also go there to be inspired by remarkable episodes of creativity and progress. And Chelsea Follett is the perfect tour guide. *Centers of Progress* is a comprehensive history lesson packed with facts yet always an enjoyable and easily accessible read."

—Johan Norberg, author of *Open: The Story of Human Progress*

"The story of human progress in all its forms is the story of civilization's greatest achievements. So, what better way to teach about those accomplishments than through a focus on the cities that were usually there at the beginning? In *Centers of Progress*, Chelsea Follett elegantly and briskly goes through most of our greatest accomplishments from the beginnings of agriculture to medicine, trade, and currency and to big ideas and events like the foundations of liberal democracy, emancipation, suffrage, and the fall of communism. And it ends, quite appropriately, with the digital revolution. To be read by all, from children to centenarians."

—John Nye, author of *War, Wine, and Taxes: The Political Economy of Anglo-French Trade, 1689–1900*

"Some times and places seem almost magical in the way they incubate ideas and movements. In explaining the magic in this fascinating book, Chelsea Follett shines a light on the drivers of human progress."

—Steven Pinker, author of *Enlightenment Now: The Case for Reason, Science, Humanism, and Progress*

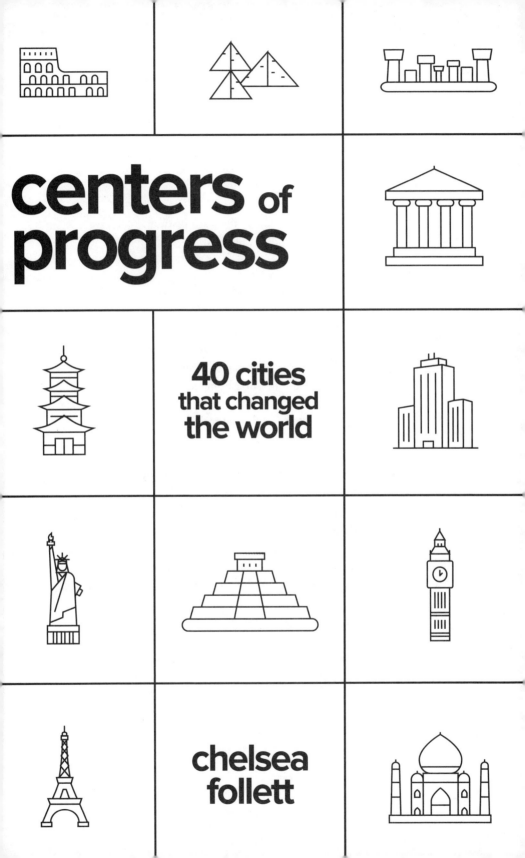

centers of progress

40 cities
that changed
the world

chelsea follett

Print ISBN: 978-1-952223-65-5
eBook ISBN: 978-1-952223-66-2

Cover and interior design: Luis Ahumada Abrigo and
Guillermina Sutter Schneider
Illustrations: Yuriy Romanovich

Library of Congress Cataloging-in-Publication Data:

Follett, Chelsea, author.
Centers of progress : 40 cities that changed the world / by Chelsea Follett.
pages   cm
Washington, DC : Cato Institute, 2023
Includes bibliographical references and index.
ISBN 9781952223655 (paperback) | ISBN 9781952223662 (ebook)
1. LCSH: Progress. 2. Social change. 3. Cities and towns--Growth.
4. Cities and towns--Social aspects. 5. Cities and towns--Political aspects.
HM891 .F65 2023
303.44--dc23/eng/20230508   2023017908

Printed in the United States of America.

CATO INSTITUTE
1000 Massachusetts Ave. NW
Washington, DC 20001
www.cato.org

To my children, Miranda and William,
the centers of my world.

Centers of Progress

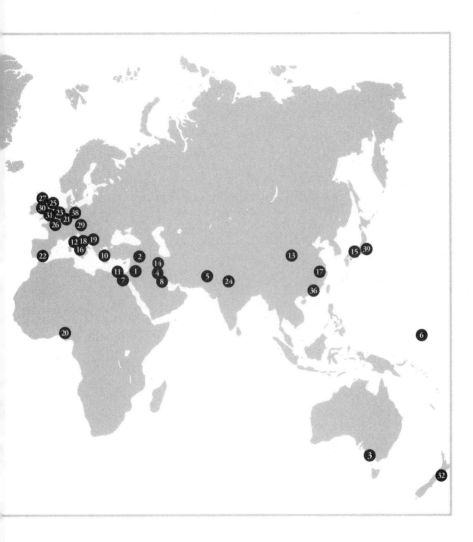

Contents

Foreword

The story of human progress is inseparable from the story of cities. From Jericho to San Francisco, from Florence to Hong Kong, innovation has happened overwhelmingly in places where people gather to truck, barter, and exchange, to borrow Adam Smith's words.

For most of human history, cities have been places of disease and violence, crime and poverty, yet they have also been crucibles of invention, places where new tools and new rules, new materials and new ideas, new art and new science were developed.

In this superb book, Chelsea Follett takes the reader on a time-travel cruise through the great flash points of human activity to catch innovations that have transformed human lives as they happened, starting in Jericho more than 11,000 years ago, at the end of the last ice age, and ending in San Francisco with the digital revolution. Sanitation was invented in ancient Pakistan, ball games in ancient Mexico, fish farming in ancient Australia, the library in ancient Alexandria, printing in 15th-century Mainz. To drop in on Bologna as a bunch of students and scholars are organizing a protective guild, or "university," or to eavesdrop on the merchants of Hangzhou as they develop the first paper currency is thrilling.

Again and again history teaches the lesson that trade is indispensable to the cross-fertilization of ideas and objects that fuel human progress. None of the centers of progress in this book were isolated from trade or contact with foreigners: quite the reverse. It is the recombination of ideas—their sexual reproduction, in effect—that drives the changes in human society.

Yet what strikes me about each of Follett's stories of progress is how transitory they proved to be. A city like Dubrovnik may find itself a global pioneer, propelled by its free-trading prosperity to lead the world in inventions—around public health in the 14th century in this case—but never for long. Manchester changed the world with industrialization in the 19th century, but its moment as a global leader lasted only a few decades. Tokyo was the epicenter of technology in the 1980s, but already it has lost its crown. Global progress depends on a series of sudden bush fires of innovation, bursting into life in unpredictable places, burning fiercely, and then dying rapidly.

Not only are the moments where progress happens almost fleeting in the great sweep of history, but the places themselves are small. Cities like Athens or Dubrovnik or Florence, even the San Francisco Bay Area that spawned Silicon Valley, are mere specks on the map of the world. Most centers of progress are city-states, those key units of civilization, rarely national capitals, and almost never great nations, let alone empires.

Indeed, empires tend to put a brake on progress, becoming steadily more suspicious of innovation, even paranoid, as the generations pass. Their centralized bureaucracies exist more and more to resist change rather than encourage it, and to enforce uniformity rather than allow diversity. The Roman, the Mayan, the Arab, the Ming, the Ottoman, the British empires (yes, even the European Union) exemplify this trend.

The exceptions seem to support the rule: the land ruled over by the Song dynasty in China a thousand years ago, which came close to sparking the Industrial Revolution seven centuries earlier than Britain did, was for most of its time hardly an empire in practice, more a confederation of self-governing city-states run by merchants. Arguably the same is true of the United States, its dominant role in the progress of the 20th century being the result of its federal structure, creating—as the name suggests—just the right balance between local experimentation and borderless free movement of goods and people.

So it is that despite inhabiting six continents on a vast globe, human beings can harvest the miraculous and cumulative improvements of human progress from these tiny, transient plots of land and moments in time. That makes human progress seem vulnerable. What would the world be like if the merchants of Dubrovnik or Amsterdam or Hangzhou had not existed or the cities had been sacked? Or, to put it another way, how many great moments of progress were lost because some ruthless warlord, such as Tamerlane or Genghis Khan or Hernán Cortés or Mao Zedong or Adolf Hitler, stamped out a promising little fire of innovation?

Yet some have argued that the invention of the computer was made possible, or at least speeded up, by World War II and the needs of the military. I think that's probably wrong. In 1937, Konrad Zuse built a prototype of a calculator in Berlin that could read a program from a punched tape. He was on the brink of something special. The same year, as Walter Isaacson has chronicled, Alan Turing published his groundbreaking paper "On Computable Numbers with an Application to the Entscheidungsproblem," John Vincent Atanasoff outlined key features of an electronic computer, Claude Shannon explained how to embody Boolean algebra in a series of switches, George Stibitz at Bell Labs designed an electrical calculator, and Howard Aiken and John Mauchly were about to get busy designing early prototypes of computers in Philadelphia and Boston.

Yet over the next few years, war drove these experts apart in secrecy so that not even the various Americans were able to put their heads together, let alone Zuse and Turing. Arguably World War II therefore delayed the progress of the computer for a decade rather than accelerating it.

Progress is a team sport, not an individual pursuit. It is a collaborative, collective thing, done between brains more than inside them. It depends on freedom, openness, and communication. Chelsea Follett's valuable book reinforces this lesson; her histories of how it happened in the past point the way to a better future for all humanity.

—Matt Ridley
May 2023

Introduction

Some people deny progress. Others find the idea of progress naive. True enough, progress has often been episodic and uneven. Furthermore, it is not inevitable or irreversible. Still, in spite of many existing problems, progress is real. To appreciate progress, step back and consider a broad perspective. The vast majority of our ancestors lived in dire poverty. Today, material abundance is more widespread than those ancestors could have ever imagined. People are living longer, healthier lives. More people than ever take adequate nutrition and proper sanitation for granted. Literacy and internet access are at all-time highs. As poverty declines, opportunities to enjoy art, music, and culture have increased. And there has been moral progress too. Slavery and torture, normalized since time immemorial, are today universally reviled. More people around the world enjoy political freedom than was the case 50 years ago.

Where did all this progress come from? Where does progress happen? The story of civilization is in many ways the story of the city. It is cities that have helped create and define the modern world by acting as the sites of pivotal advances in culture, politics, science, technology, and more. Or as the physicist Geoffrey West put it, "Cities are the crucible of civilization." This book explores 40 examples of exactly that.

There is a reason why progress tends to emerge from cities, even though the vast majority of humans who have ever lived resided in rural areas. Rural life should not be romanticized. Of course, rural communities have made plenty of achievements as well. From the early humans in Kenya's Great Rift Valley who pioneered stone-tool use to the Wright brothers from the rural Midwest who first took flight in an empty North Carolina field, people outside cities have also contributed to human progress. But living in a sparsely populated community means fewer choices; fewer people to work with, befriend, or marry; fewer choices on where to work, shop, relax, eat, and worship—the list goes on.

Cities are gathering places. And wherever many people meet face-to-face, their potential to accomplish amazing things increases. The great evolutionary advantage of *Homo sapiens* is not physical strength or speed. Humans are a relatively puny species, compared with apex predators. But our problem-solving abilities, which are magnified when we work together, are second to none.

Cities are marketplaces, hubs of commerce. Cities are factories, centers of production. Cities are hotbeds of creativity, where artists compete and collaborate. Cities are laboratories, sites of experimentation. Cities are great big classrooms, where we debate and learn from one another. But those things are only true under certain conditions.

Although there are some exceptions, most cities reach their creative peak during periods of peace. Most centers of progress also thrive during times of relative social, intellectual, and economic freedom, as well as openness to intercultural exchange and trade. And centers of progress tend to be highly populated. Not all the places featured in this book are, strictly speaking, cities by our modern standards. But in the historical context, all represent significant population centers. That makes sense because, in every city, ultimately the people who live there drive progress forward—if given the freedom to do so.

City life also has its challenges, to be sure. When people live close together, they are more vulnerable to dangers such as nuclear strikes and pandemics. And because technology now allows people to work together productively even while thousands of miles apart, continued progress need not necessarily rely on ever-higher levels of urbanization. If more people choose to live in quieter areas while working remotely, humanity may reach a turning point where the long-standing trend of urbanization reverses. But whatever the future may hold, the role that cities have played in the history of progress is worth understanding.

It is mind-boggling that the world has improved in all the ways it has. But history is too often told as a story of degeneration. Many intellectuals and everyday people alike embrace this "declension narrative" of history. They see history as a lengthy tale of decay since some lost, idealized golden age in the past, while dismissing the overwhelming evidence of progress.

This book is both a popular and a dissident history. It is "popular" in the sense of being easy to read and meant for a general audience, not just academics. It is "dissident" because, unlike many academic historians, it takes seriously the idea of progress. That wasn't always the case. From the later 18th century onward, plenty of historians believed in progress, but often thought about progress in an ethnocentric way. They could see that progress was real, but they credited that progress to specific religions, races, or nationalities without much thought about the policies or institutions that economists now believe are central to the story of progress.

That old view of progress is now often described as propaganda to hide the excesses of imperialism and instances of exploitation that accompanied widespread advances. The ascendant view is that every culture is just as good as every other, and no cultural innovations are any better than any others. That's an inaccurate story too. It is certainly possible for one flawed narrative to be replaced by another. And that's what has happened.

There is no question that certain places, at certain times in history, have contributed disproportionately toward making the world a better place. The desire to tell the story of those places inspired me to write the *Centers of Progress* series of articles that originally appeared on HumanProgress.org and became this book. Progress is not the exclusive destiny of people belonging to any one creed, race, or nation. The centers of progress featured in this book include a diverse set of societies, ranging from ancient Uruk to Song-era Hangzhou.

But some common themes stand out. Again, those are (relative) peace, freedom, and multitudes of people. Identifying those common denominators among the places that have produced history's greatest achievements is one way to learn what causes progress in the first place. After all, change is a constant, but progress is not. Understanding what makes it possible for a place to achieve progress may help stimulate future progress.

Perhaps the biggest reason why cities produce so much progress is that city dwellers have often enjoyed more freedom than their rural counterparts. Medieval serfs fleeing feudal lands to gain freedom in cities inspired the German saying *stadtluft macht frei* (city air makes you free). That adage referred to laws granting serfs independence after a year and a day of urban residency. But the phrase has a wider application than its coiners intended.

Cities have often been havens for innovators throughout history—and indeed for anyone stifled by the stricter norms common in smaller communities. The journalist H. L. Mencken once opined, "Human progress is furthered, not by conformity, but by aberration."

City air not only makes one free. City air also provides the wind in the sails of the modern world: a world of increasing wealth, health, knowledge, creativity, and—most important—freedom. Come journey through these pages to some of history's greatest centers of progress.

1

Jericho

AGRICULTURE

Our first center of progress is Jericho. Jericho is thought by many scholars to be the world's oldest city. It was first settled sometime around 9000 BCE. The people who lived in Jericho and the surrounding areas may have been among the first humans to give up their hunter-gatherer ways, to domesticate plants, and to become farmers.

The invention of agriculture, often called the First Agricultural Revolution or the Neolithic Revolution, was a decisive turning point in our species' history. It dramatically changed the way that we live. By producing a surplus of food that could be stored for difficult times ahead or traded for other goods, agriculture ultimately allowed for far greater prosperity than hunting and gathering ever could.

Today, Jericho is a tourist-oriented city in the Jordan River Valley and is frequented by religious pilgrims and history buffs.

It is relatively small, with a population of just over 20,000 people. The city is located in a natural oasis in the desert, thus earning its nickname in the Hebrew Bible—the City of Palm Trees. The city is home to various cafés selling shawarma and falafel, as well as to many historic ruins. Jericho is also the site of near-constant archaeological digs, as we try to deepen our knowledge of the city's past.

If you were to visit Neolithic Jericho, you may have been able to observe two decisive events in the history of civilization: permanent settlement and the beginnings of agriculture.

Imagine a group of hunter-gatherers—dubbed "Natufians" by today's archaeologists—walking through the wilderness. They would have carried hunting weapons such as spears, and they would have worn leather made from the hides of mountain gazelles and beaded jewelry made of gazelle bones. They would have carried food and supplies in baskets and animal skins. They would also have had domesticated dogs walking alongside them, perhaps looking something like the modern-day basenji.

You would have seen them coming upon a natural oasis bursting with freshwater springs in the middle of the wilderness and settling down to rest. You would have watched this group of hunter-gatherers coming to a momentous decision as they resolved—perhaps after some spirited discussion in a long-dead language—to build a permanent camp at the oasis and end their nomadic wanderings.

Of course, the decision was probably gradual, with the Natufians camping out at the oasis for longer and longer periods each year, until the settlement became their home year-round. But at some point, the decision was made to remain there permanently. In any case, the Natufians built a number of semisubterranean oval-shaped stone dwellings to form a village that would grow into the world's first city. Thus, the story of Jericho began.

The first people to inhabit what would become Jericho had long survived by hunting animals such as gazelles and eating wild cereals and other wild plants. But a shift in the climate, which became less rainy and more desertlike, may have helped prompt a change in the Natufian survival strategy.

How did that happen? Maybe the Natufians noticed that edible plants sprouted in places where those plants' seeds had been scattered before. Perhaps inspired by that observation, an enterprising individual (or multiple individuals) must have, at some point, proposed deliberately planting the seeds of the plants that the Natufians ate. When the Natufians began to plant seeds intentionally, they set humanity on a new course.

The Natufians are often called the first farmers. Although there is no expert consensus on precisely where in the Fertile Crescent agriculture first appeared, Jericho was certainly among the earliest communities to practice agriculture. The oldest archaeological remains of domesticated barley, rye, and early forms of wheat are found in human Neolithic sites in the Fertile Crescent, such as the Natufian settlement where Jericho is today. Evidence of cultivated figs has also been found near Jericho dating to about 9400 BCE.

The world's first farmers were patient and innovative. Consider wheat. They discovered how to selectively breed wild emmer grass so that the plant's seeds would not fall off its stalks when the grass became ripe, making collection of the seeds far easier. They used the seeds to make bread, and what started as just another kind of grass gradually became what we now know as wheat. Today, according to research by Rachel Brenchley of England's University of Salford and several coauthors in the scientific journal *Nature*, 20 percent of the world's total calorie consumption comes from wheat.

Researchers disagree on how much credit ought to be given to the conscious efforts of the early farmers. "One controversy in this area is about the extent to which ancient peoples knew they were domesticating crops," noted University of Sheffield plant scientist Colin Osborne. "Did they know they were breeding domestication characteristics into crops, or did these characteristics just evolve as the first farmers sowed wild plants into cultivated soil, and tended and harvested them?"

In addition to bread, the Natufians also enjoyed beer. And some researchers believe that the production of alcoholic beverages made from fermented cereals may have served as one of the motivations underlying early agriculture.

Whatever their motivation, the first Jerichoans became farmers and were thus able to produce enough food to eventually leave their old hunter-gatherer lifestyle behind. Selectively breeding plants would prove to be a painstakingly slow process, and perhaps for centuries, the people of Jericho may have continued to supplement their agricultural food production with hunting and gathering.

As agriculture advanced, the people still hunted gazelles and other game animals, but the grains they planted, harvested, and stored yearly increased their food security. One day, there was no longer any need to forage for wild plants, beginning a new chapter in human history.

Over the centuries, the residents of Jericho became increasingly adept at farming. The Jerichoans went on to cultivate many other plants and to develop an irrigation system, and their harvests grew larger. They soon had enough food to store for lean times and to trade. But with such productivity came a danger—the threat that nearby nomadic tribes would raid the city and rob Jericho's granaries with their large stores of food.

To fend off raiders, the people of Jericho built the oldest known protective wall in the world, perhaps dating to about 8000 BCE. At that point, Jericho's population had probably reached 2,000 people or so. For perspective, that's about as many people as the current population of the rural town of Victor, Idaho. For its time, however, Jericho must have felt like a bustling metropolis. (Recall that fewer than 10 million people lived in the entire world back then, roughly equivalent to the current population of Portugal.)

Producing a surplus of food allowed for some specialization of economic activities: not everyone had to be a farmer, freeing people to pursue other projects. The wall's construction could not have been accomplished without some degree of specialization. The stone wall stood over 11 feet high; in addition to defending the city, it may have also served to protect the city from floods.

Some evidence suggests the wall's accompanying 28-foot-high cone-shaped stone tower, also built about 8000 BCE, served a symbolic purpose rather than a practical one. The tower is not well positioned to serve as a defensive lookout. But computer models show that when the tower was built, the nearby mountains would cast a shadow on it just as the sun set on the longest day of the year—the summer solstice. The shadow would fall precisely on the tower and then spread out to cover all of ancient Jericho.

So the tower may have served as a warning: its growing shadow let the people of Jericho know that the days ahead were about to become shorter and the nights longer. Agricultural activities such as planting and harvesting are intimately linked to different seasons of the year, and so, to the largely agrarian community of Jericho, marking the summer solstice likely held great significance. The solstice may have been observed as a day of importance, whether as a celebratory festival or a day of solemnity.

The tower may also have symbolized power or authority. Transitioning from hunting and gathering to becoming an agricultural society entailed a transformation in how people related to one another: whereas hunter-gatherer tribes tended to be egalitarian (i.e., lacking in hierarchy), the more specialized and complex society that emerged in Jericho brought with it a new set of social power dynamics. Grave sites show that the first Jerichoans observed differences in rank, with some individuals buried alongside valuable goods such as shell jewelry and others occupying simpler graves.

"This was a time when hierarchy began and leadership was established," according to Tel Aviv University archaeologist Ran Barkai, one of the researchers behind the discovery of the connection between the Tower of Jericho and the summer solstice. "We believe this tower [by acting as a symbol of power and authority] was one of the mechanisms to motivate people to take part in a communal lifestyle."

Today, Jericho is perhaps best known for the role it played in biblical times. It is the place the ancient Israelites are purported to have conquered in 1400 BCE, after escaping from slavery in Egypt. The well-known song about the Battle of Jericho—covered by musical icons ranging from Bing Crosby to Elvis Presley—was first composed in the 19th-century United States by enslaved African Americans. The song's subject, about a previously enslaved people triumphant in battle, and its chorus, proclaiming that Jericho's "walls came tumbling down," both alluded to the songwriters' own desire for freedom.

Thus, the city of Jericho became a symbol of freedom in popular culture many thousands of years after that city helped free humanity from foraging for food in the wilderness. The transition to agriculture was likely a difficult and patience-testing process, upending the Natufians' previous way of life and altering their social structure, but the payoff has been a level of food security beyond what our hunter-gatherer ancestors could have imagined.

For being the world's oldest city and possibly the birthplace of agriculture, Neolithic-era Jericho deserves to be recognized as our first center of progress.

Göbekli Tepe

RELIGION

Our next center of progress is Göbekli Tepe, the site containing the oldest-known monumental structures and perhaps the earliest archaeological evidence of religious practice. Although much disagreement exists on the origins of religion, many scholars describe Göbekli Tepe as the world's first manmade temple, sanctuary, or holy place. Göbekli Tepe serves as a reminder of humanity's capacity to create impressive structures, as well as the long history of systems of faith and their profound influence on the world.

Göbekli Tepe lies in the southeast of what is today Turkey, about 30 miles from the border with Syria. Today, only a small portion of the prehistoric site of worship has been excavated, and much of it likely remains buried underground. Göbekli

Tepe consists of large, ringed enclosures measuring as wide as 65 feet across, as well as rectangular pillar arrangements that may have once supported roofs. Each ring is made up of over 40 T-shaped stone pillars, some as tall as 18 feet. Another 250 or so pillars may remain buried. Some of the uncovered pillars are blank, but many feature detailed totem-like carvings depicting people, abstract symbols, and a wide variety of animals, such as foxes, lions, bulls, scorpions, snakes, wild boars, birds, spiders, and insects. Some carvings appear to be part human and part animal and may represent deities. The pillars are the oldest-known megaliths, predating the better-known Stonehenge by millennia.

Boardwalks now encircle the main excavation site, allowing tourists to view the pillars from different angles. And a roof has been constructed over the stones to protect the carvings and archaeologists from the sweltering sun. In July, the average temperature in the area exceeds 100 degrees Fahrenheit. Although the climate is only classified as semidesert, rain almost never falls during the summer.

But if you could visit Göbekli Tepe in its heyday, you would encounter a very different world. The climate was wetter, and the surrounding environment was a vast grassland filled with wild goats and gazelles. Looking out over the endless fields, you would see tall grasses, such as einkorn, wheat, and barley, rippling in the wind. Rivers and waterfowl may have been visible as well. Your view of the surrounding plateaus would be excellent, as Göbekli Tepe stands on top of a hill. The name Göbekli Tepe, in fact, means "potbelly hill" in Turkish.

Radiocarbon dating suggests the currently exposed structures of Göbekli Tepe were built over centuries, with some parts perhaps dating to 9600 BCE and others constructed as recently as 8000 BCE or even 7000 BCE. It was a time of significant change. Communities like the ancient Natufians of Neolithic-era Jericho, located 500 miles southwest of Göbekli Tepe, were making the momentous transition from nomadic hunting and gathering to permanent settlement and agriculture. The people who built Göbekli Tepe may have still been mainly hunter-gatherers, but they also likely farmed in villages for at least part of the year. Archaeological evidence shows that their diets consisted largely of meat but were supplemented by cereals that they probably farmed.

The religious site may have even helped inspire agriculture. According to author Sean Thomas:

> Gobekli Tepe upends our view of human history. We always thought that agriculture came first, then civilisation: farming, pottery, social hierarchies. But here it is reversed, it seems the ritual center came first, then when enough hunter gathering people collected to worship [they invented agriculture].

Erecting and carving humanity's earliest monuments was a painstaking undertaking that required a multigenerational investment of time, labor, and craftsmanship. It likely involved hundreds of men. The people who built Göbekli Tepe did not yet have pottery or metal tools, or the help of domesticated animals or wheeled vehicles. Flint tools would have been sufficient to carve the pillars, made of relatively soft limestone.

There is no proof that anyone ever lived at Göbekli Tepe, although some scholars believe it was nonetheless a settlement. Whether the site offered sufficient access to water to sustain residents is much debated, and a lack of trash pit remnants suggests that people did not sleep at the site. Perhaps only a single person (such as a priest or shaman) or a small number of people resided there, leaving no archaeological footprint that has yet been discovered. But even though Göbekli Tepe's builders may have camped elsewhere, the site was certainly alive with activity. It may have been the closest thing to an urban center that the nomadic hunter-gatherers knew.

Turning away from the magnificent grasslands toward the imposing structures of Göbekli Tepe, one would have been struck by the aroma of freshly roasted wild boar, gazelle, red deer, and duck and witnessed the local hunter-gatherers commencing a festival amid their monuments. Researchers believe the hunter-gatherers congregated at the site to dance, celebrate, drink beer made from fermented grains, and dine together. In addition to food preparation tools, archaeologists have so far uncovered about 650 carved stone platters and vessels at the site, some large enough to hold over 50 gallons of liquid. More than 100,000 bone fragments from wild game also suggest feasting. Such ritual feasts may have originated sometime between 8000 BCE and 6000 BCE, when the transition to agriculture linked the relative scarcity or abundance of food to certain seasons of the year. Among the festivities held at Göbekli Tepe may have been "work feasts" held throughout the site's multigenerational construction to celebrate the completion of different sections of the temple.

From the Passover seders of Judaism to the Eid al-Fitr (nicknamed Sugar Feast) sweets of Islam, and from the Christmas dinners of Christianity to the staple desserts of Hinduism's Diwali, religious feasts continue to hold great importance to communities across the world.

Much remains unknown about the nature of Göbekli Tepe and the religion that may have inspired its establishment. Prominent vulture carvings at the site have led some scholars to conclude that the religion was a "funerary cult" centered on venerating the dead. However, no human remains have been uncovered to suggest that Göbekli Tepe was ever a cemetery. Others think that the site was linked to astronomy and that its carvings reference constellations and comets. Some believe that Göbekli Tepe was a temple to the brightest star in Earth's night sky, Sirius, because the central pillars may have framed the star as it rose. The main archaeological team excavating the site rejects claims of an astronomical link, though.

Some scholars also think Göbekli Tepe may have been a holy site attracting hunter-gatherer visitors from across the Levant and as far away as Africa. Knowledge of the site would have traveled by word of mouth, since writing did not exist yet. According to journalist Charles Mann:

> Göbekli Tepe may have been the destination for a religious pilgrimage, a monument for spiritual travelers to be awed by a religious experience—like the travel now made by pilgrims to the Vatican, Mecca, Jerusalem, Bodh Gaya (where Buddha was enlightened), or Cahokia (the enormous Native American complex near St. Louis).

Objects found at the site support this theory. Researchers have traced certain obsidian artifacts to volcanoes hundreds of miles away, and other tools found among the ruins exhibit carving styles suggesting far-flung origins, such as the eastern Mediterranean. However, these objects could have also come to Göbekli Tepe via trade between different hunter-gatherer bands. Göbekli Tepe represented "a very cosmopolitan area . . . almost the nodal point of the Near East," claims University of Toronto anthropologist Tristan Carter. "In theory, you could have people with different languages, very different cultures, coming together."

At some point, the Neolithic people decided to bury Göbekli Tepe. Maybe their religion changed, and the site lost its relevance to them, or maybe the burial was itself a ritual tied to their particular spiritual beliefs. The site's remarkable level of preservation is credited to the way in which it was buried. The hunter-gatherers then built another layer of stone pillars on top of the buried temple.

Religious faiths continue to provide a sense of meaning, structure, guidance, and inner peace to many people today— about 84 percent of people globally, according to the Pew Research Center. Although the negative effects of violent strains of religious extremism are undeniable and religious conflict has caused much suffering, faith has also uplifted humanity in many ways.

In fact, religious inspiration is a common factor among several of the centers of progress featured in this book. Some scholars think the religion of the ancient Indus valley civilization may have been based on cleanliness, helping incentivize Mohenjo-daro's achievements in sanitation. In Baghdad, during that city's golden age, the then-prevailing interpretation of Islam helped motivate scientific inquiry and the pursuit of knowledge. In Renaissance Florence, faith inspired many leading artists, and the Catholic Church funded groundbreaking artistic projects. During the Scottish Enlightenment that birthed modern social science, the dominant moderate branch of the Presbyterian Church embraced cutting-edge thinkers in Edinburgh. And later, prominent Anglican clerics supported London's trailblazing quest to end the global slave trade. In each of these cases, religion encouraged some manner of positive innovation.

That is not to downplay the harms that can arise from highly illiberal forms of religion. Examples include the restrictive interpretation of Islam that ultimately contributed to unraveling Baghdad's status as a center of learning or the extremist Christian movement led by the radical friar Girolamo Savonarola that sought to destroy Florence's artworks.

Happily, liberty-minded thinkers can be found among the adherents of all major religions today. See, for example, the scholarship of Mustafa Akyol on the Muslim case for liberty, the writings of Stephanie Slade on the Catholic case for liberty, the words of Russ Roberts on the Jewish case for liberty, and the work of Aaron Ross Powell on the Buddhist case for liberty. And for an ecumenical Christian case for the same, see the works of the Acton Institute for the Study of Religion and Liberty. Their writings illustrate how faith can champion the freedom needed to discover and create remarkable things.

Although we may never learn why Göbekli Tepe was built, the site's megalithic structures and intricate carvings arguably symbolize the power of religious devotion. The sophistication and artistic achievement embodied by this creation of a largely preagricultural society are astounding. If the site indeed served as a gathering place where prehistoric people worshiped now long-forgotten deities together, then it stands as a testament to the many ways in which humanity has sought to understand our place in the universe and express reverence. The mysterious, gigantic Stone Age site is worthy of being called a center of progress.

3

Budj Bim
AQUACULTURE

The next center of progress is the historic site at Budj Bim in southeastern Australia. Budj Bim, meaning "high head," is a dormant volcano, the dried lava of which has been crafted into a series of manmade channels, weirs, walls, and dams that may represent humanity's oldest aquaculture system.

Aquaculture is the cultivation of aquatic organisms. Some forms of aquaculture, like fish and eel farming, involve animal husbandry, a breakthrough in food security. Animals, after all, are harder to manage than immobile plants but are also a better source of protein. Aquaculture shaped early human society in some areas of the world as much as agriculture did in others, encouraging permanent settlement and defining the rhythms of daily life. The vast aquaculture complex at Budj Bim exemplifies the innovative ways in which humans have shaped their physical environments to combat hunger throughout history.

A few other animals cultivate food from watery surroundings. The damselfish, for example, weeds its rudimentary algae gardens and aggressively defends the "crop" from other much larger creatures. However, no other living creature besides humans has come anywhere close to true aquaculture.

The ruins at Budj Bim are older than the Egyptian pyramids and the English Stonehenge. Parts of the stonework system were built before 4500 BCE, predating early examples of hydraulic engineering in many Northern Hemisphere civilizations. Radiocarbon dating suggests that humans may have created many of the system's artificial ponds as far back as 6000 BCE. Construction of some of the extensive site's groundwork may have even begun between 7000 and 6000 BCE.

Today, this large area of modified wetland, spanning about 40 square miles, is peaceful and remote: a tranquil scene of water, volcanic rock, and wildlife. Picnickers enjoy the view as black swans glide along the many spring-fed creeks, and koalas look on from above in the tall, twisted manna gum trees and angular blackwoods. Many areas that were underwater when the aquaculture system was active are now dry. But evidence of the locale's ancient significance can be seen in the scattered stone remnants of prehistoric eel traps, manmade channels, and house sites spread across the Budj Bim area. A recent series of wildfires revealed previously unknown swaths of the complex that had been overgrown by vegetation.

The native people are known as Gunditjmara, an Aboriginal Australian clan group. In 2019, UNESCO designated the Budj Bim Cultural Landscape as a World Heritage site, noting, "Aquaculture acted as the economic and social base for Gunditjmara society for [at least] six millennia." Of course, it is possible that other clan groups also contributed to the creation and maintenance of the stone complex at Budj Bim over the course of its lengthy history.

The site was probably born out of a series of volcanic eruptions that began about 30,000 years ago, which created the outpouring of lava that hardened into basaltic rock and later provided the raw building material for the aquaculture system. The Budj Bim volcano, also known as Mount Eccles, erupted at least 10 times, with the most recent eruption occurring about 7,000 years ago, or about 5000 BCE. Stone tools found underneath the oldest layer of volcanic ash prove that humans have inhabited the area since before the volcano erupted. The Gunditjmara's oral histories seem to describe the volcano's erupting as part of a creation myth, which geologists Erin Matchan, David Phillips, Fred Jourdan, and Korien Oostingh take as evidence that the Gunditjmara may have "some of the oldest oral traditions in existence." The Gunditjmara pride themselves on their tradition of storytelling. According to their mythology, the now-dormant volcano is a creator god or ancestral being that brought Gunditjmara society into existence. The Gunditjmara call the rock-filled lava flow area *tungatt mirring* (stone country).

It is certainly true that the hardened lava from the volcano provided an advantageous natural resource. Ultimately, though, it was human ingenuity that transformed the lava landforms and waterways from a rocky swamp into a steady source of abundant food. Eel farms provided the mainstay of the Gunditjmara diet and a product to trade with other clan groups. Aquaculture, in other words, furnished the basic driver of their economy and culture. The practice was also intertwined with the Gunditjmara religion, and they considered the eel to be a sacred animal. The people also farmed galaxias fish and ate freshwater mussels and other aquatic creatures. They further supplemented their seafood diet with the meat of land animals they hunted, such as ducks, as well as plains turkeys, goannas, and kangaroos. They managed their hunting grounds with a system of low-intensity, intentional fires that burned away hazardous dry brush and helped create ideal habitats for hunting game. They also cultivated and ate various vegetables like murnong, also called yam daisy.

As with agriculture, the tasks required to maintain an aquaculture-based society are often dictated by the changing seasons. Whereas some eels can be found in the area year-round, during certain periods of the year, they number in the millions. The native eel species, *Anguilla australis*, can grow to over 40 inches long and weigh more than seven pounds. The local galaxias fish—a slim species with a mottled pattern, usually about four inches long—are also migratory and, in the right season, could be caught in the tens of thousands. In the spring, the eels travel along rivers from the sea to their marshy inland feeding grounds. During the subsequent wet season, the marshlands burst with eels. Then, in autumn, the eels return to the sea to breed.

The local people recognized that these predictable patterns of migration provided an opportunity that they could exploit to ensure a stable supply of food. "It shows us they had a high level of technical skill, understanding of physics and of the natural environment," according to University of Washington archaeologist Ben Marwick. Drawing on their observations of changes in water levels and eel migration routes, the Gunditjmara people manipulated the seasonal flooding with manmade channels and weirs, diverting the water flow to trap eels and fish. They also took care not to overharvest and risk depleting the eel or fish populations.

If you could travel back in time to when the aquaculture system was in active use, you might observe workers carefully adjusting the stonework, perhaps replacing the basalt in an area where older stones had fallen away or adding in a new section. Marwick believes the ancient engineers "continually modified the system." The stones formed a complex network of artificial channels—some over 1,000 feet long—that diverted water to shepherd migrating eels and fish. Some of the aquatic creatures were driven into handwoven nets for immediate harvest, and others were guided into holding ponds or pens to be collected later. All in all, there were at least 70 functional aquaculture systems. In those artificial ponds, the corralled eels grew fat, feeding on local insects, water snails, frogs, and small fish, until the time came for them to be eaten. Woven baskets set in weirs built from volcanic rock and wooden lattice structures would then capture the seafood.

Walking away from the elaborate trap system to visit the nearby settled community of perhaps 600 people—although that population estimate is likely to be revised upward—built on the edge of the waterways, you would have seen numerous stone huts with fireplaces. You would have also seen women weaving baskets for the weirs used to cull mature eels, men returning from the eel traps hauling a fresh harvest in such baskets, and people preparing the eels for consumption, first by cleaning and gutting them. And you would have witnessed them smoking the rich, oily eel meat with burning leaves from blackwood trees. Researchers have found eel lipids in the earth beneath burnt, hollowed-out trees, suggesting that they were used as family cooking hearths and smokehouses to prepare the eels for trade with other tribes.

Smoking is often considered humanity's earliest method of meat preservation, allowing meat to be stored for the off-season, as well as transported and used as a trading commodity. By drying the flesh, smoking makes it less susceptible to bacteria that need moisture to grow, and chemicals released from the smoke have antibacterial properties that further safeguard the meat. Smoking also cooks the eel meat, which is poisonous when raw. Eel blood contains a potentially deadly toxin that cramps muscles and can stop the heart from beating. Cooking breaks down the toxin. The Gunditjmara served the eels in a variety of ways. The eels' bones and skin could be used to create flavorful cooking stock, and the meat could be complemented with local plants, such as kelp and saltbush.

For millennia, the aquaculture system yielded a reliable food supply, and it was still in use when the British came to the area in the 19th century and provided the first written accounts of the elaborate stone-walled facilities. In 1841, the British colonial official and preacher George Augustus Robinson arrived on an exploratory expedition and described the aquaculture system as "resembling the work of civilized man." But he also noted, evidencing the prejudices of the era, that "on inspection I found [it] to be the work of the Aboriginal natives, and constructed for the purpose of catching eels."

"It is hardly possible for a single fish to escape," he continued. "Triple water courses led to other ramified and extensive trenches of a most torturous form."

Today, the Gunditjmara people comanage, with the Australian government, the Budj Bim National Park, which encompasses the ruins of the sprawling Budj Bim aquaculture system. Some of the descendants of the ancient engineers and fishermen who masterminded the aquaculture complex still catch eels and cook them using traditional methods. Various Australian localities even hold eel festivals celebrating eel recipes, both ancient and modern.

A steady supply of food is necessary for any society to function. Budj Bim demonstrates the antiquity of humanity's quest to stave off hunger by deliberately managing the environment. For millennia, the Gunditjmara transformed and enriched their local ecosystems with clearing fires, stone infrastructure, and artificial ponds. Their elaborate system of water manipulation to systematically trap, store, and harvest seafood represents one of the oldest aquaculture systems in the world. For those reasons, Budj Bim is fittingly a center of progress.

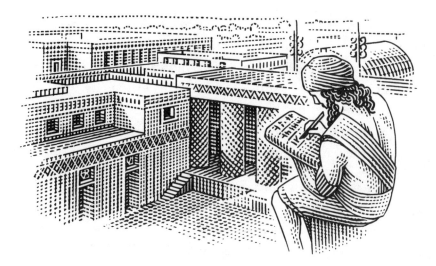

4

Uruk

WRITING

Our next center of progress is Uruk, the world's first large city and the birthplace of writing circa 3300 BCE. By creating the first writing system, the people of Uruk revolutionized humanity's ability to exchange information.

Before the invention of writing, the main way people could communicate was by speaking to each other in person. Communication over vast distances and across long stretches of time was restricted by the fallibility of human memory. It was possible to send a messenger to a faraway city, but there was always a risk that the messenger would not arrive safely or would not recite the message accurately. People could pass down knowledge and histories through oral traditions from one generation to the next, but the details tended to change over time.

Today, Uruk is an uninhabited archaeological site preserved in the desert of southern Iraq. It is part of a UNESCO World Heritage site, honoring the "relict landscape of the Mesopotamian cities." You can still see the remains of the city walls and gates, make out the shape of the streets and the layout of the houses from their crumbling foundations, and view the cracked steps of the temple mounds.

Today's Uruk is quiet and ghostly. But if you were to visit Uruk in the late fourth millennium BCE, you would have entered a thriving hub of art and commerce populated by about 10,000 inhabitants. That would increase to between 30,000 and 50,000 inhabitants by the beginning of the third millennium BCE.

For perspective, Uruk's population in the late fourth millennium BCE was about the same as the population of the small town of Brattleboro, Vermont, today. But Uruk was among the first settlements to achieve a population of that size and is considered by many to be the world's first large city. In the year 3200 BCE, Uruk was the largest city in Mesopotamia and possibly in the entire world.

As Uruk's population grew, its society became more complex and the Sumerian civilization (the world's first true civilization, which flourished in southern Mesopotamia between 4500 and 1500 BCE) reached its creative peak. Surviving tablets indicate that Uruk had over 100 different professions, including ambassadors, priests, stonecutters, gardeners, weavers, smiths, cooks, jewelers, and potters.

Walking through the streets of Bronze Age Uruk, you would have seen merchants hawking their wares, beautiful gardens with palm trees, and temples rising high above all of the other structures. The temple complexes were places of religious importance, but that was not their only purpose. You may have seen men carrying clay pitchers filled with grain into the temples, because these imposing monuments were also where the people of Uruk stored their surplus food.

The arid desert around Uruk had few natural resources. To compensate, the people developed robust trade networks with other communities. They imported wood from the Taurus, Zagros, and Lebanon Mountains, and lapis lazuli gemstones from as far away as what is today known as Afghanistan. Some of these valuable imports were also stored in the temples.

Near one of the temple's entryways, you may have witnessed a history-altering breakthrough. You may have seen an accountant or record keeper marking a clay tablet each time a pitcher of grain entered the temple. He would have made a small picture of a grain stalk next to his tally marks, like the city's record keepers had done for centuries.

But if you looked more closely, you would have observed that his picture was not really a picture at all. That is because, over the course of many years, the record keepers' pictures had become simpler to make taking inventory of goods faster. Eventually, the image that was used to represent grain in the temple records no longer even vaguely resembled a grain stalk. The pictographs evolved, in other words, to become nonpictorial symbols that represented concepts—such as grain.

By agreeing on a set of abstract symbols to represent common goods stored in their temple warehouses, Uruk's accountants could avoid the laborious chore of making detailed drawings on their clay tablets.

Eventually, the people of Uruk used these written symbols not only to represent different concepts, like grain or fish or sheep, but also to represent the spoken sounds that people used to express those concepts. Once they had symbols for different sounds, it became possible to write names or other words phonetically. After that innovation, the Sumerians could write down more than simple inventory lists. They could also create increasingly complex documents. Their written output ranged from lengthy epic poems and wisdom literature to genealogies and lists of kings.

According to the writings of the ancient Sumerians, the city of Uruk was built by the mythical king Enmerkar. This epic hero was thought to have been the son of the Sumerian sun god Utu and a cow (an animal that the Sumerians revered and associated with motherhood because of its production of milk). Enmerkar is said to have ruled Uruk for hundreds of years. If the mythical figure of Enmerkar is loosely based on a real ruler, then he would have lived at the end of the fourth or beginning of the third millennium BCE.

In Sumerian legend, which was preserved in the epic *Enmerkar and the Lord of Aratta*, Enmerkar is credited with the invention of writing. The legend says that he did so during a period of tense negotiation with a neighboring king, the ruler of the rival city-state Aratta. Enmerkar was purportedly dissatisfied that his messenger—who was exhausted from traveling back and forth between Uruk and Aratta reciting messages—could only relay messages of limited length to convey to the neighboring king.

So Enmerkar purportedly picked up some clay, magically created a complete written language, and proceeded to write down a message for his messenger to deliver to the king of Aratta. Specifically, the myth states:

> [King Enmerkar's] speech was substantial, and its contents extensive. The messenger, whose mouth was heavy, was not able to repeat it. Because the messenger, whose mouth was tired, was not able to repeat it, the lord of [Uruk] patted some clay and wrote the message as if on a tablet. Formerly, the writing of messages on clay was not established. Now, under that sun and on that day, it was indeed so.

That colorful legend shows that the Sumerians valued written language so highly they thought only a king (and an ostensible demigod, no less) could create something so important.

In reality, writing was not invented by a king, but by the city's accountants. Moreover, it was not created all at once in a burst of creative genius, but arose gradually over many generations. It was not originally created to gain an advantage in international diplomacy, but instead came about for the far less glamorous reason mentioned earlier: bookkeeping. As such, the earliest writings that survive today tend to be inventory lists, shopping lists, wage records, lists allocating rations for temple workers, and shopping receipts.

The people of Uruk wrote with reeds and clay, because those materials were both widely available. Uruk is situated amid the Mesopotamian Marshes, a rare watery landscape in the middle of an otherwise dry desert. The wetlands, fed by the Euphrates and Tigris Rivers, may have been larger in the past than they are today. A channel of the Euphrates that has since dried up is thought to have flowed very close to Uruk.

After cutting a reed from the marshy banks of the Euphrates, the people of Uruk at some point discovered that when a single reed is pressed, with its cut edge facing down into soft wet clay, it produces a distinctive wedge shape. When the clay dried and hardened, that shape was preserved.

When the accountants simplified their pictographs into ever more abstract symbols, those symbols took the form of certain arrangements of wedge-shaped marks, which then became the first characters or "letters." That is why the earliest writing system is now known as cuneiform, from the Latin for "wedge-shaped."

Originally, record keepers would take inventory by writing from top to bottom on their clay tablets, as if making a list. After many years of writing that way, the scribes developed an innovative new system of writing from left to right. That innovation reduced the risk of smudging what had been written before the clay dried.

However, the temple priests and other literate people of Uruk were accustomed to reading records from top to bottom, not from left to right, and did not care for the scribes' new system. The scribes came up with a solution that would allow them to write from left to right, while still allowing their tablets to be read from top to bottom. Ingeniously, the scribes simply wrote down versions of their written symbols that were rotated 90 degrees. Writing their symbols sideways allowed those reading the tablets in the old way, from top to bottom, not to be inconvenienced.

Eventually, people began to read the symbolic script in the same way that it was written, from left to right. But because the already abstract symbols were rotated, they became even more abstract, hastening the process of moving from simple pictographs to cuneiform characters. Below is the evolution of the cuneiform character for "head," from a simple picture drawn circa 3000 BCE into a highly abstract cuneiform character almost 1,000 years later.

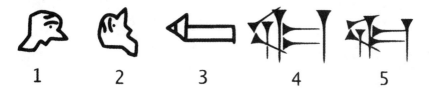

Today, Uruk is best known as the city of the ancient hero Gilgamesh, as described in the *Epic of Gilgamesh*. That epic poem began as a series of poems composed circa 2100 BCE, although the most complete surviving version is considerably more recent, dating to the 12th century BCE.

Scholars believe that a real person named Gilgamesh probably ruled Uruk sometime between 2800 and 2500 BCE, and came to be described as a demigod and larger-than-life hero after his death. Thanks to the invention of writing, people today are able to enjoy not only Sumerian literature, but all of human literary output ranging from the plays of William Shakespeare to the science fiction of Isaac Asimov.

For being the world's first large city and the birthplace of writing, Bronze Age Uruk deserves to be recognized as a center of progress. Writing gave humanity a new means of creative self-expression and the ability to exchange information across generations and across the globe.

5

Mohenjo-Daro
SANITATION

Our next center of progress is Mohenjo-daro, a city in today's Pakistan that pioneered new standards of urban sanitation. The city is thought to have been constructed circa 2500 BCE, although the site has been inhabited since circa 3500 BCE. Mohenjo-daro was the largest urban center of the ancient Indus valley civilization, covering over 600 acres, and one of the world's earliest major cities.

The people of the Indus valley civilization invented water supply and sanitation devices that were the first of their kind. They included piping and a complex sewage system. Tunnels under Mohenjo-daro carried the city's waste to a nearby estuary. Almost all the city's houses had indoor baths and latrines with drains, and the city also showed its dedication to cleanliness with a large public bathhouse used for ritual bathing. In *National Geographic*, writer Traci Watson has opined that that civilization enjoyed the "ancient world's best plumbing," in some ways surpassing even the plumbing system that the Roman civilization would later create.

Ever since humanity gave up hunting and gathering to live in permanent settlements, our species has faced health challenges related to hygiene and the proper disposal of waste. Since the advent of cities, humanity has been vulnerable to rapidly spreading illnesses, because disease propagates more easily in concentrated populations. That is particularly true without adequate sanitation, and waterborne illnesses—such as cholera, dysentery, hepatitis A, typhoid, and various gastrointestinal diseases—were once a common cause of death.

Advances in sanitation have allowed people to live near one another in cities with less risk to their health than in the past. In particular, safe disposal of effluent to spare the water supply from contamination has proved to be a truly game-changing innovation. It has been argued that plumbers are the unsung heroes of civilization.

Today, Mohenjo-daro is a striking archaeological site in Pakistan's Sindh Province. The site's name means "mound of the dead" in Sindhi. Only part of the ancient city has been excavated and much of it remains hidden. Mohenjo-daro has been designated as a UNESCO World Heritage site. Located on the right bank of the Indus River, Mohenjo-daro is the most impressive of the ruined cities remaining from the Indus valley civilization. Mohenjo-daro's surviving structures are made of bricks fashioned from red sand, clay, and stones, lending the ruins a ruddy hue.

The Indus valley civilization arose in the floodplains of the Indus and Sarasvati Rivers in what is now northwestern India and Pakistan, around 5,000 years ago. (Note: the latter river presents a bit of a historical puzzle and may not have existed.) The rivers flooded twice a year in a predictable manner, making the land fertile and allowing the Indus people to farm everything from cotton to dates to support their growing population.

Their prosperity also flowed from conflict avoidance and from their vast trade networks. They established one of the first long-distance trade relationships in the world by exchanging goods with the Mesopotamians located nearly 2,000 miles to the west, as early as 3000 BCE. Indus exports included spices such as clove heads, luxury goods like carnelian beads artfully etched with acid, and possibly even livestock such as water buffalo. Their imports from the Mesopotamians included textiles and various artistic motifs and legends—including aspects of the legend that would come to be known as the *Epic of Gilgamesh*. The Indus people also had what is thought to be a written language, now called the Indus script, which has yet to be deciphered by scholars.

If you could have visited Mohenjo-daro in its heyday, you would have seen an orderly city of dense, multistoried homes with flat roofs fashioned out of uniformly sized bricks, standing along a grid of perpendicular streets. The grander houses had up to 12 rooms. You would have seen people gathering water, in decorated pottery jugs, from the numerous public wells and chatting, perhaps discussing art. Archaeological evidence suggests that Mohenjo-daro's residents enjoyed art forms ranging from metal sculpture to dance. You may have observed children playing games, including games with dice, which many historians believe the Indus people invented.

The city's population may have peaked at about 40,000 people, similar to the population of Annapolis, Maryland, today. The men probably wore a cloth around their waists, perhaps gathered in a way that resembles the modern dhoti, while the women wore longer skirts or robes. Wealthy people of both sexes wore jewelry with ivory, lapis lazuli, carnelian, and gold beads, as well as elaborate hairstyles and headdresses.

Walking through the city, you would have observed that Mohenjo-daro had no grand temples, palaces, monuments, or royal tombs. The society of Mohenjo-daro seems to have been far less hierarchical than the cities of the Mesopotamians with which the former traded. The people in Mohenjo-daro may have had no king or, if they had one, he had only little authority. The lack of any royal structures certainly suggests the absence of a powerful ruler, although it remains unknown what kind of system governed the city.

Instead of a palace, the largest structure in the city was an immense, elevated public bathhouse. The Great Bath of Mohenjo-daro measured almost 900 square feet, with a maximum depth of around 8 feet. It was constructed of fine brickwork, with a pool floor made of three layers: sawed brick set in gypsum mortar, then a bitumen sealer, followed by another layer of sawed brick and gypsum mortar.

The status of the bathhouse as the city's biggest and most prominent structure suggests the people of Mohenjo-daro highly valued cleanliness. Their entire ideology may have been based on cleanliness, according to anthropological archaeologist Gregory Possehl of the University of Pennsylvania.

The bathhouse may have been a sacred place, and scholars believe it was likely used for ritual bathing. The people did not need to use the bathhouse for everyday washing, because practically every home in the city—from the largest to the humblest—had what was then a remarkable, groundbreaking feature: a washroom.

These rooms were typically small, and square or rectangular in shape. In each washroom, the brick-paved floor was carefully built to slope toward a corner containing a simple latrine and drain, as well as a drained washing area. The slanted floors helped ensure proper drainage, and the bricks were set tightly together to prevent leaking. Around each drain hole, the bricks were so meticulously rubbed down and fitted together that the joints were nearly invisible. In some cases, the bricks were overlaid on a bed of pottery debris to further bolster the floor's resistance to leaks.

The Indus were the first people to have indoor plumbing, perhaps as early as 3000 BCE. Homes with washrooms on upper floors were fitted with vertical terra-cotta pipes that carried effluent down to the street level. The pipes of fired clay were joined with tar to make them watertight. The pipes were positioned so that wastewater flowed down into the drainage ditches that ran along every avenue in the city, and then into underground tunnels. Thanks to the invention of drainage ditches, the cleanliness of the city's streets was remarkable for the ancient world.

As the city's population grew and the amount of waste it processed increased, the people kept their drainage ditches functional by raising the brick walls alongside the former—to prevent effluent overflow into the streets. Archaeological evidence suggests the walls grew gradually in size to meet the city's needs. The ditches and connected underground sewage tunnels carried waste away from the city, protecting its well-water supply from contamination.

Just like modern washrooms, the washrooms of Mohenjo-daro were used for multiple personal hygiene activities, including bathing. Surviving artifacts suggest the Indus poured pottery jugs of water over themselves to shower, and used clay scrapers similar to Greco-Roman strigils to cleanse themselves. In these rooms, they also used pottery rasps to remove cuticles and shape their nails. Some washroom ruins contain what may be oil residue, suggesting the washroom was also where Mohenjo-daro's residents moisturized their skin with oils.

Some traditions appear to be timeless. For example, evidence suggests that Mohenjo-daro's children played with bath toys just like today's children. Instead of rubber ducks and plastic boats, their toy figurines were made of pottery. "To judge from the number of pottery models that have been found in the drains, it would seem that the childish habit of taking play-things into the bath has persisted for thousands of years," according to British archaeologist Ernest Mackay, who led the excavation of Mohenjo-daro in the 1920s and 1930s.

Children were arguably the greatest beneficiaries of Mohenjo-daro's dedication to hygiene, although the city's washrooms and sewage system were essential to the health of all of its people. Although it can be hard to imagine for those of us lucky enough to be able to take modern sanitation for granted, standards of hygiene throughout most of human history have been appalling. Associated illnesses were responsible for high rates of mortality, especially among children.

Mohenjo-daro's advanced plumbing serves as a reminder that progress is not steady or linear. Many people who lived thousands of years later coped with conditions far less hygienic than those enjoyed by Mohenjo-daro's people in the third millennium BCE.

It was not until the 19th century that urban sanitation became widespread. Those advances—along with the discovery of the germ theory of disease—are the primary reasons for the dramatic rise in human life expectancy, according to Nobel Prize–winning economist Angus Deaton. Although more people now enjoy proper sanitation than at any other time in history, even today, in poor areas of the world, far too many people contend with inadequate sanitation and accompanying diseases.

Mohenjo-daro is thought to have been gradually abandoned almost 4,000 years ago, when the Indus River shifted its course and farmers could no longer rely on it to irrigate their crops. Today, Mohenjo-daro is best known as the largest remnant of the enigmatic Indus valley civilization. Because the Indus people's writing system is currently unreadable, many aspects of that civilization remain a mystery. The religion and seemingly kingless government system of Mohenjo-daro are unknown, as are the reasons for the Indus valley civilization's ultimate demise.

For developing plumbing and wastewater management, Mohenjo-daro has earned its place as a center of progress. Without washrooms and sewage systems, our lives would be far shorter and less hygienic.

6

Nan Madol

SEAFARING

Our next center of progress is Nan Madol, a city that showcases the far reach of the first seafarers. Micronesia, where Nan Madol was eventually built, started to be settled by the ancient Austronesians over 4,000 years ago. They are the people who are thought to be the first humans to invent seafaring vessels.

Their invention allowed them to explore and populate much of the Indo-Pacific region during the so-called Austronesian expansion. The expansion peaked between 3000 and 1000 BCE—although the Austronesians did not reach some islands in Polynesia until after the year 1000 CE and may not have settled New Zealand until around 1300.

The stone city of Nan Madol—constructed circa 1100 (with parts perhaps dating to as early as the year 500), toward the end of the Austronesian age of discovery—stands as a testament to the first seafarers' ingenuity and the wide reach of their exploration.

Before the advent of seafaring vessels, people could not travel beyond coastlines. As such, many hospitable lands remained uninhabited by humans because no one could reach them. Various cultures independently created dugout canoes to travel along rivers, but the oceans remained impassable. Eventually, people came to wonder what lay farther away—across the oceans. The invention of seafaring boats allowed the ancient Austronesians to explore new lands and quite literally expanded humanity's horizons.

The invention of seaworthy vessels was likely a gradual process of improving on riverboats until they could handle the rough waters of the open ocean. To people back then, the ocean must have seemed as daunting an obstacle as outer space seems to us, but they persevered. Some of the early attempts at voyaging were no doubt failures, resulting in many lives tragically lost at sea. But each time the explorers successfully ventured out a bit farther from the shore and returned safely, their confidence grew.

The first true oceangoing vessels were watercraft featuring lateral support floats known as outriggers, which are secured to one or both sides of the boat's main hull. The outriggers help stabilize the boat and prevent it from capsizing on the choppy waters of the open sea. The first outriggers may have been simple logs or fallen tree branches, but their shape was refined over the years with careful craftsmanship to maximize stability. To steer their outrigger vessels, the first seafarers often used sails woven from salt-resistant pandanus leaves.

In due course, the seafarers developed catamarans, or watercraft featuring parallel hulls in place of mere outriggers.

Some catamarans were large enough to carry more than 80 people and could handle being out at sea for months on end.

Today, the ruined stone city of Nan Madol stands on elevated, artificial islets on the eastern edge of the island of Pohnpei, which is a little smaller in area than New York City. The island is now part of the Federated States of Micronesia. Nan Madol is a designated UNESCO World Heritage site. Nan Madol means "in the space between things," in reference to the canals crisscrossing between the islets. The stone city's remains have been partially overgrown by mangroves and other plants, lending the ruins an eerie aura. The well-known horror fiction writer H. P. Lovecraft was inspired by the ruins and modeled the city where his famed fictional monster Cthulhu dwells after Nan Madol.

Scholars disagree on precisely where the Austronesian people originated, with theories ranging from Taiwan to various islands in Southeast Asia. But whatever their point of origin, the Austronesians' outrigger canoes and catamarans allowed them to spread out across the Pacific, exploring and settling new lands. Before the 16th-century European Age of Discovery, Austronesians were the most widespread ethnolinguistic group, spanning half the planet. Remnants of Austronesian settlements can be found in places as disparate as New Zealand in Oceania, Easter Island in the southeastern Pacific Ocean near South America, and Madagascar in Africa.

Even as they spread out across half the world over the course of centuries, the vastly dispersed Austronesian population maintained many things in common. They spoke variations of the same language and shared many of the same technologies and traditions, such as body tattooing, jade carving, megalithic construction, stilt houses, and various artistic motifs.

The Austronesians also partook in a common set of agricultural techniques and raised many of the same animals, such as chickens, pigs, and dogs, and grew many of the same plants. Those included bananas, coconuts, breadfruit, yams, and taro. They transported seeds and animals on their boats during their sea migrations. After their settlements reached their greatest extent east, they introduced sweet potatoes from South America to the islands of the Pacific and to Southeast Asia circa 1000–1100 CE.

Among the places that the Austronesians settled during their expansion were the Micronesian islands located in the western Pacific Ocean. They were the first humans to set foot there, as the remote islands were unreachable before the invention of seaworthy vessels. It is believed the eastern Micronesian islands such as Pohnpei were first settled by seafarers venturing out from the islands now known as Vanuatu and Fiji, sometime before 1000 BCE. Archaeological and linguistic evidence suggests the settlers worked their way up the chain of islands gradually.

During the peak of the Austronesian expansion, despite building a variety of impressive megalithic stone artworks— such as those in Lore Lindu National Park on the island of Sulawesi in Indonesia, which perhaps date to circa 2000 BCE— the Austronesian people mainly constructed their homes of materials that decompose. As a result, they left no well-preserved ruined cities from that era.

Sometime around the year 1100 CE, the seafarers found a reason to build a city out of a more lasting material: stone.

The decentralized chieftain-based system of Micronesia had changed into a more unified economic and religious culture centered on Pohnpei and the island's chief. That chief, who established the Saudeleur dynasty and a system of absolute rule, chose to build a royal and ceremonial city out of stone for the prestige that the material conferred. The ruins of Nan Madol have thereby survived through the centuries and offer us a window into the Austronesians' lives. Evidence suggests Nan Madol was the first capital city in the Pacific representing unity under a single chief.

The city's walls are built of basalt cut into the shape of interlocking logs—perhaps a legacy of the wood-based building techniques the culture practiced before shifting to stone as a building material. The walls enclose an area almost 5,000 feet long and nearly 2,000 feet wide. The total weight of the basalt columns or logs that had to be transported to construct the city is estimated at 750,000 metric tons.

That means that to build Nan Madol, the people of Pohnpei moved an average of 1,850 tons per year over four centuries. Given that the island had a population of fewer than 30,000 people, that was a great feat. The method they used to move the stones remains a mystery. "Not bad for people who had no pulleys, no levers and no metal," noted Rufino Mauricio—an archaeologist who works at the Nan Madol site and who is affiliated with the Federated States of Micronesia's Office of National Archives, Culture, and Historic Preservation.

If you could visit Nan Madol in its prime, you would be struck by the way boats acted as a vital part of the city. The city contains nearly 100 artificial manmade islets or platforms built of stone and coral that are crisscrossed by tidal canals. Nan Madol's canals served as the thoroughfares on which people traveled around the city, using wooden canoes. The city has thus been nicknamed the Venice of the Pacific. It is the only ancient city ever to be built atop a coral reef.

Raising your gaze beyond the canals, you would have seen an affluent city containing stone palaces, temples, mortuaries, and residential homes for the society's nobility. The city was constructed in part to serve as an enclave to house the chief and the nobility. You would thus have immediately noticed that the majority of the roughly 1,000 residents were either high-status individuals, extensively adorned in jewelry such as necklaces and arm rings, or their servants. The walls surrounding the city acted as fortifications to protect the high-status inhabitants.

But Nan Madol was also home to a sprawling urban marketplace, in which you could purchase goods ranging from trochus shells and quartz crystals to distinctive pottery. You would have seen merchants hawking trolling lures for fishing, stone and shell tools, and carefully carved bead necklaces. Evidence suggests the marketplace may also have sold pet dogs and turtles. Food stalls likely sold pork, poultry, fish, rice, copra, bananas, breadfruit, and taro—among other foods.

No food was grown in Nan Madol itself. Instead, the city served as a center of trade for food and other goods transported by boat from elsewhere in the chiefdom. Nan Madol was also a major spiritual center and the site of many religious ceremonies, such as an annual ceremony of atonement in which priests would offer cooked turtle to saltwater eels. (Eels were considered to be sacred.) Additionally, Nan Madol was a place for diplomacy and important political gatherings.

After expanding their territory to cover half the world, the Austronesians mysteriously ceased voyaging, instead settling down permanently in places such as Nan Madol. According to legend, Nan Madol was founded by twin brothers who came to Pohnpei by boat from an unknown island, seeking a place to build an altar so that they could worship the god of agriculture. That legend reflects the end of the age of voyaging and the transition to a stationary, agricultural way of life.

Today, Nan Madol is best known as the seat of power during the Saudeleur dynasty of chiefs, who ruled from circa 1100 to circa 1628. According to the oral histories passed down on Pohnpei, the dynasty became increasingly oppressive with each generation, as each new chief sought to replace the autonomous and decentralized Pohnpeian culture with an ever more abusive and centralized system of social control. The rule of the last chief was so cruel that it sparked mass discontent.

The oral histories relate that the dynasty ended when that tyrant chief was overthrown (with the full support of the local people) by Isokelekel—a semimythical hero and warrior. He was purported to be a demigod from the neighboring Micronesian island of Kosrae and came to power in the 16th or 17th century. His successors abandoned Nan Madol in the early 19th century.

The invention of sea travel was a revolutionary breakthrough in the history of transportation. By showing the far reach of the first seafarers, Nan Madol has earned its place as a center of progress. The same spirit of discovery that drove people to set sail on the ocean in search of new lands eventually guided our species to develop air travel and space travel, and may one day lead humanity to set foot on other planets.

7

Memphis

MEDICINE

The next center of progress is Memphis, an important center and capital of ancient Egypt that greatly advanced humanity's understanding of medicine. The ancient Egyptians pioneered medical specialization and arguably invented rational (nonmagical) medicine.

Memphis is the Greek or Hellenized name for the city, which the Egyptians called Men-nefer (possibly meaning "beautiful harbor") since at least the third millennium BCE. Today, the archaeological zone at Memphis is a designated UNESCO World Heritage site. Tourists flock to view what is left of the ancient city in the Memphis Open Air Museum, which includes a famed fallen 30-foot-tall limestone colossus of Ramses II, Egypt's most powerful ruler, who reigned from 1279 to 1213 BCE. Outside the museum, visitors throng around other excavated monuments, such as a giant alabaster sphinx and another large statue of Ramses II made of granite. The nearby necropolis at Saqqara, home to Egypt's oldest pyramid and the tombs of numerous pharaohs, also draws many visitors—indeed, the English historian John Julius Norwich has called Memphis "a city unusually overshadowed by its cemeteries."

Memphis is located south of the Nile River delta, about 15 miles from modern Cairo, right at the entrance to the Nile River valley. This strategic location perhaps destined the site to become the nucleus of Egyptian commerce and the capital of Lower Egypt, an independent political entity that existed from circa 3500 to circa 3100 BCE in the northernmost region of Egypt.

Archaeological evidence of agriculture and animal domestication suggests the area has been inhabited since Neolithic times and had a well-developed culture by circa 3600 BCE. According to tradition, however, the city was founded in 2925 BCE by Menes, Egypt's semimythical first pharaoh, who is credited with uniting the prehistoric kingdoms of Upper and Lower Egypt and thus establishing the Egyptian state.

The Greek historian Herodotus claimed that Menes drained the plain of Memphis and built a large dam around the city to shield it from the sometimes-catastrophic Nile floods. Some scholars believe the name Menes may mean "the Memphite," further linking Egypt's founding to the city of Memphis. Menes is said to have reigned for 62 years before a lethal wildlife encounter. Conflicting accounts suggest he was killed by a hippopotamus, a crocodile, or an allergic reaction to a wasp sting (if only the EpiPen had been invented, he might have lived).

Memphis was not only united Egypt's first capital but served as the capital on and off "for the best part of three and a half millennia, from the beginning of the pharaonic period (c. 3000 BCE) until the Arab conquest (641 CE)," according to Norwich. During the Second Dynasty (c. 2890–c. 2686 BCE), the capital moved to Thinis (the capital of Upper Egypt before unification with the north). But Memphis again served as Egypt's capital city for the Third, Fourth, Fifth, Sixth, Seventh, and Eighth Dynasties. Even after the seat of government moved to Thebes (c. 2240 BCE), Memphis remained one of Egypt's chief cultural, religious, and economic centers for centuries.

During the Old Kingdom (c. 2700–c. 2200 BCE), Egypt's first golden age, Memphis was home to as many as 30,000 people, making it perhaps the world's largest settlement at the time. If you could have visited the thriving, palm-filled city, you would have observed administrators, workers, and slaves (like most ancient societies, Egypt engaged in the practice of slavery) walking to and from the palace, people haggling for goods in the marketplace or chatting while playing board games, and worshipers thronging around the many temples. Within those temples, the ill could seek treatment in medical institutions called the Houses of Life, which were established in Memphis as early as the First Dynasty (c. 3100–c. 2900 BCE).

Urban centers have often been at the forefront of medical progress. One of the world's oldest systems of medicine—which even included cosmetic reconstructive surgery—originated in the ancient Indian city of Kashi. Rossi Thomson, writing for the BBC, calls the Italian city of Padua, home to the first permanent anatomical theater, "the birthplace of modern medicine." The first successful heart surgery, another turning point in medical history, took place in Cape Town, South Africa. Many point to Greece as the cradle of Western medicine, and the medical profession certainly owes a debt of gratitude to the Greek healer Hippocrates. But Memphis deserves distinct credit, as it was home to perhaps the earliest pioneer of medicine.

That great medical innovator was Imhotep, the chief minister and head magician for the Third Dynasty pharaoh Djoser (reigned c. 2650–c. 2575 BCE), whose court was in Memphis. The *Encyclopedia Britannica* names Imhotep the "first physician." Imhotep is also thought to have designed the world's oldest stone pyramid, the step pyramid built at the necropolis of Saqqara outside Memphis, which houses Djoser's tomb. Some think Imhotep founded the oldest school of medicine in Memphis.

In those days, the *Encyclopedia Britannica* notes, "the chief magician of the pharaoh's court also frequently served as the nation's chief physician," underscoring the blurred boundary between magic and medicine throughout much of human history. Yet an ancient Egyptian document known as the Edwin Smith surgical papyrus—dating to circa 1600 BCE but written in archaic hieroglyphs thought to be copied from a much older papyrus sometimes attributed to Imhotep—may represent the oldest known case of rational (nonmagical) medicine. (The document takes its name from an American collector of antiquities who possessed it at one point.) It's a straightforward surgical guide and may have been a military medical field manual.

Humanity's earliest efforts to treat disease were often highly unscientific, relying on rituals, incantations, and other literal attempts to work magic. Some ancient Babylonians thought kissing a human skull seven times before bed could cure night-time teeth grinding, and some ancient Romans thought that consuming the blood of fallen gladiators could cure epilepsy. "Abracadabra," the famed gibberish incantation, was once a supposed treatment for malaria. In the second century CE, the Roman writer Serenus Sammonicus in his *Liber Medicinalis* (*Medical Book*) advised fevered patients to write the magic word "abracadabra" over and over on a piece of paper, tie the paper with flax, wear it as a necklace for nine days and then, before sunrise, toss the charm into an east-running stream. Throughout much of antiquity, illness prompted a visit not from a doctor but a shaman healer or magician.

The oldest known medical procedure was quite extreme. Between 5 and 10 percent of skulls from the Neolithic era show evidence of trepanation: the deliberate drilling or scraping of holes in the skull, probably in an attempt to treat epilepsy, mental illness, or head injuries. Strangely, that primitive surgery may represent a precursor to rational medicine. Prehistoric people likely observed that head injuries resulted in the loss of consciousness more often than other injuries and concluded that the head held special significance. "The head was chosen for the procedure, not because of . . . magic . . . but because of . . . accumulated experience observed by primitive man in the Stone Age with ubiquitous head injuries during altercations and hunting," according to Cuban American medical historian Miguel Faria.

There may have been a certain logic to trepanation; however, the often-fatal surgery is no longer in use, for good reason. The ancient Egyptians almost never employed the technique, although they made impressive breakthroughs in surgery. The first recorded account of a surgical suture dates to circa 3000 BCE in Egypt, and the oldest confirmed suture is on an Egyptian mummy thought to date to circa 1100 BCE. The ancient Egyptian physicians created sutures from plant fibers, tendons, hair, and woolen threads.

The ancient Egyptians were arguably the first people to develop a medical system with a high level of documentation, and a growing body of research suggests that rational medicine existed in Egypt before Greece. The Greeks themselves admired Egyptian medicine. Homer (c. 800 BCE) remarked in the *Odyssey*, "In Egypt, the men are more skilled in medicine than any of humankind." Hippocrates, Herophilos, Erasistratus, and later Galen studied in Egypt and acknowledged the Egyptian influence on Greek medicine. Italian medical scholar Marco Rossi even claims that "Egyptian medicine is the base of Greek medicine."

The Edwin Smith papyrus has been called "the birth of analytical thinking in medicine." It describes 48 different medical scenarios, mostly involving traumatic injuries. The text instructs the physician on how to examine the patient, describes the outlook for the patient's survival based on the physical signs revealed by the examination, and suggests specific treatments, including simple surgeries. We now take physical examinations, diagnoses, and prognoses for granted; however, at the time, these were extraordinary breakthroughs.

Another well-preserved medical treatise, the Ebers papyrus (c. 1550 BCE, but likely copied from an older text), consisted of both "magical" treatments and more sensible remedies. The document is named after the 19th-century German Egyptologist Georg Ebers. (The course of action the Ebers papyrus recommends for Guinea worm disease—wrapping the emerging end of the worm around a stick and slowly pulling it out—remains the standard treatment to this day.) Rational medicine coexisted with magic-based medical practices for millennia and would not start to supplant the latter until the Scientific Revolution in the 16th and 17th centuries. The medical system developed in Memphis—like the much-lauded Greek medicine that came later with its emphasis on "humorism," bloodletting, and belief in "wandering wombs"—included many bizarre errors.

Although no modern patient would want to receive treatment in Memphis, the medical advancements made there were remarkable for the era. Ancient Egypt was arguably the birthplace of anatomical science, partly thanks to the practice of mummification—a method of preserving the body after death—which dates to circa 3500 BCE and was already entrenched in Egyptian society by the time Memphis rose to prominence. Embalmers and physicians in Memphis broke new ground in the understanding of the circulatory system and internal organs and knew how to take a pulse.

The physicians of Memphis also developed many medical specializations. According to Herodotus, "The practice of medicine [was] so specialized among [Egyptians] that each physician [was] a healer of one disease and no more." Egyptian writings refer to "eye physicians," "stomach physicians," "shepherds of the anus" (i.e., proctologists), and more. Many scholars believe one of the titles of Hesy-Ra, a high court official in Memphis during the early Third Dynasty (c. 2650–c. 2575 BCE), may translate to "great one of the dentists"—making him the first dentist whose name is known to history. The Kahun gynecological papyrus (c. 1825 BCE) represents the oldest surviving text on gynecology. A noblewoman named Peseshet, who lived circa

2500 BCE, when Memphis was the capital, held the title "lady overseer of women physicians" and may be the earliest named female doctor.

In addition to specialization, the physicians of ancient Egypt discovered many effective treatments and pioneered such areas as surgery, nutrition, pharmacology, and prosthetics. For example, University of Manchester Egyptologist Ann Rosalie David noted in *The Lancet*, "Anatomical and radiological studies on skeletal and mummified remains [have] revealed healed fractures and amputation sites, confirming that the Egyptians did successful surgery." A healed mandible suggests successful oral surgery as early as the Fourth Dynasty (c. 2575–c. 2465 BCE). In Memphis, court surgeons used scalpels made from copper, ivory, or obsidian. And perhaps as early as 3000 BCE, the Egyptians made a medicinal drink from boiled willow tree bark for pain relief. Centuries later, the active ingredient salicin formed the basis of the discovery of aspirin, which remains one of the most commonly used drugs in the world.

A century after Imhotep's death, Egyptians began to worship him as the god of healing. Posthumous deification was a rare honor for nonroyal Egyptians, but Imhotep's cult grew over the centuries until he became one of the patron deities of Memphis. In the Ptolemaic times, Memphis's importance dwindled as the new seaport of Alexandria (another center of progress highlighted in this book) supplanted the former as an intellectual center, exporting Egyptian medical understanding to other parts of the Mediterranean. The Arab conquest in the seventh century CE dealt the last blow to Memphis, as the city became a quarry, stripped for building materials to construct new settlements, including Fustat (now Cairo's city center), the Arab capital.

Without rational medicine, medical specialists, and the numerous other foundational advances in the treatment of disease originating in ancient Memphis, our lives would be far shorter and sicklier. It is for those reasons that Memphis deserves its place as a center of progress.

Ur

LAW

Our next center of progress is the Mesopotamian city of Ur during the so-called Sumerian Renaissance, in the 21st century BCE. Ur then served as the capital city of a king named Ur-Nammu. Under his direction, the city issued the oldest surviving legal code in the world, the Code of Ur-Nammu, which predates the better-known Code of Hammurabi by three centuries. Ur-Nammu's code of laws, which were carved onto terra-cotta tablets and distributed throughout his kingdom, represented a significant breakthrough in the history of human civilization.

The Code of Ur-Nammu helped establish the idea of a set punishment for a particular crime that applied equally to all free persons regardless of their wealth or status. In other words, the code replaced arbitrary standards of justice, which shifted with each new instance of a crime, with a uniform and transparent set of rules. Many of those rules were horrific by modern standards, but the code nonetheless represented a notable development toward what we now consider to be the rule of law.

References in ancient Sumerian poetry suggest the existence of an even older legal code than the Code of Ur-Nammu, called the Code of Urukagina, written in the 24th century BCE. Unfortunately, the text of that earlier code has not survived. The Code of Ur-Nammu, as the oldest surviving legal code, is thus the best window that we have into the origins of lawmaking.

Today, the city of Ur lies in ruins in the desert of southern Iraq. Ur's Great Ziggurat, erected to honor the Sumerian god of the moon, still stands. The Ur archaeological site is also home to what may be the oldest still-standing arch in the world. Many of the artifacts found at Ur have been relocated and can now be seen in the British Museum in London and the University of Pennsylvania Museum of Archaeology in Philadelphia. Ur is part of a UNESCO World Heritage site that also includes Uruk, which you may recall as another center of progress, which is less than 60 miles away.

During its golden age, Ur was the capital of a state including all of Babylonia and several territories to its east. It was also a key port of trade between Babylonia and regions to the south and east.

Picture the city, surrounded by palm trees and skillfully irrigated land, made fertile by tributary streams flowing to the river Euphrates that lay to the west. As you approached, you would have seen farmers tending barley fields, fishermen casting their nets into the streams, and herdsmen leading their sheep to graze.

As you entered the bustling urban center itself, you would have observed its many people. Ur's population eventually swelled to 65,000. That may not seem like a lot—it is roughly the same as the modern-day population of Lynchburg, Virginia—but it was about 0.1 percent of the entire global population at the time. Ur would become the most populous city in the world and remain so until circa 1980 BCE.

The people of Ur wore skirts or wraps of *kaunakes*, a woolen fabric with a tufted pattern like overlapping leaves or petals. The rich wore belts of gold or silver, and wealthy women wore hair ornaments and jewelry of the same materials. Everyone, even royalty, went barefoot. Sandals would not appear in the region until centuries later. The city natives mainly had dark hair— the people of Sumer referred to themselves as the "black-headed ones." The people of Ur likely shared the city streets with oxen pulling along wagons heaped with supplies, and the stench of manure may have been inescapable. The very wealthy traveled in chariots pulled by donkeys, or perhaps onager hybrids.

The city's architecture featured columns, arches, vaults, and domes. You would perhaps have seen people carrying baskets filled with offerings on their heads walking toward one of the city's temples to its numerous gods. The city's temples were richly decorated with statues (often with blue lapis eyes), mosaics, and metal reliefs. The temple columns were sheathed with colorful mosaics or polished copper. Inscribed tablets lay at the temples' foundations.

You would have seen the space where work had begun on a three-storied ziggurat made of mud brick faced with burnt bricks set in bitumen. Upon that platform, a temple would soon be constructed. The temple—honoring the moon god Nanna, the patron deity of Ur—would tower over the city and be visible from the far distance in the flat surrounding Mesopotamian countryside. The partially reconstructed ziggurat stands today as the most prominent structure of Ur.

At the edge of the sacred precinct was the Royal Cemetery, out of use by that time for 50 some years. There, about 1,800 people lay buried—royalty laid to rest wearing elaborate gold ornaments, alongside their attendants, victims of human sacrifice. But the city had abandoned that practice by the era that concerns us.

In the marketplace, you would have seen artisans selling their wares, such as woolen textiles, clothes, and tapestries; jars, fluted bowls, and goblets, some made of precious metals; elaborate carved stone vessels of chlorite, bearing cuneiform inscriptions; ornaments and jewelry of semiprecious stones, such as carnelian, and precious metals; and various tools and weapons. Passing through the food stalls you would likely have seen wheat, barley, lentils, beans, garlic, onions, and goat's milk. You would have seen stone vessels of wine and precious oils.

You might have paused at a stall selling carved musical instruments, stopping to admire a lyre featuring lapis lazuli—a stone all the way from the upper reaches of the Kokcha River, over 1,000 miles away in what is now Afghanistan. Its presence is a reminder of the city's far-reaching trade.

Moving along, you might have observed two men hunched over a strategy board game. The Game of Ur was then popular throughout Mesopotamia among people of all social strata. Perhaps you would have heard the players arguing about the rules, and then watched them turn to a clay tablet serving as a rule book to resolve their dispute. (Such tablets, describing the game's rules, have survived.)

The people of Ur had a guide to help them navigate disputes concerning far larger matters as well. If you were to visit in the year that the locals called the "Year Ur-Nammu made justice in the land," believed to be circa 2045 BCE, then you could have witnessed a history-altering moment. You would have perhaps had the good fortune to watch as Ur's messengers disembarked from the city to deliver tablets bearing the new legal code throughout the kingdom.

The Code of Ur-Nammu, as the oldest surviving legal code, helped redefine how people conceptualized justice. The Code of Ur-Nammu listed laws in a cause-and-effect format (i.e., "if this, then that") that specifically outlined different crimes and their respective punishments. A total of 32 laws survive.

The Code of Ur-Nammu also introduced the concept of fines as a form of punishment—a notion we still rely on today. Fines ranged from minas and shekels of silver to *kur*s of barley. (The Sumerian measurement system is not fully understood, but a *kur* or *gur* was likely a unit based on the estimated weight that a donkey could carry.)

Compared with the later Code of Hammurabi, the Code of Ur-Nammu was relatively progressive, often imposing fines rather than physical punishment on the transgressor. In other words, it often favored compensation for the crime's victim over the enactment of retributive justice against the crime's perpetrator. The Code of Hammurabi famously dictated, "If a man put out the eye of another man, his eye shall be put out."

That "an eye for an eye" rule is also cited in the Old Testament books of Exodus and Leviticus. In contrast, the older Code of Ur-Nammu states, "If a man knocks out the eye of another man, he shall [pay] half a mina of silver."

In the prologue to the code, King Ur-Nammu boasted about his various accomplishments and claimed to have established "equity in the land." By equity, he did not mean the modern concept of equality—after all, he ruled over a society with widespread slavery. But by establishing uniform punishments for crimes, he meant to ensure that both rich and poor free persons were treated equally before the law. In the prologue, he noted: "I did not deliver the orphan to the rich. I did not deliver the widow to the mighty. I did not deliver the man with but one shekel to the man with one mina [i.e., 60 shekels]. . . . I did not impose orders. I eliminated enmity, violence, and cries for justice. I established justice in the land."

The king clearly saw his legal code as an important part of his legacy and wanted to be remembered as a just ruler. The code certainly represented a step forward when compared with a purely arbitrary system of punishment. It was arguably more humane than some legal codes that followed, such as the aforementioned Code of Hammurabi. That said, the Code of Ur-Nammu is not one that a modern person would want to live under. Some of the laws were ridiculous ("If a man is accused of sorcery he must undergo ordeal by water"), sexist ("If the wife of a man followed after another man and he slept with her, they shall slay that woman, but that male shall be set free"), or plain barbaric ("If a man's slave-woman, comparing herself to her mistress, speaks insolently to her, her mouth shall be scoured with 1 quart of salt").

Some laws were also oddly specific, such as, "If someone severed the nose of another man with a copper knife, he must pay two-thirds of a mina [1.25 pounds] of silver." Was there a different punishment if the knife used was not made of copper? (Today, if you're curious, cutting off someone's nose will land you in prison for 1 to 20 years—at least in Rhode Island, the only state I could find with a law that specifically mentions nose mutilation.)

Today, the city of Ur is perhaps best known for being thought to be the birthplace of the biblical patriarch Abraham. Abraham is an important figure in the religions of Judaism, Christianity, and Islam, which are thus known as "the Abrahamic religions" for that commonality.

The advent of laws transformed how communities enact justice by ensuring a uniform and transparent set of rules. Although many laws throughout history have proved to be mistakes, and unjust laws continue to pose serious problems in many countries, a system of laws is nonetheless better than a system where punishments are doled out without any consistency and at the whim of a ruler or a mob. By enacting the oldest surviving legal code, Sumerian Renaissance–era Ur has earned its place as a center of progress.

9
Chichén Itzá
TEAM SPORTS

Our next center of progress is the Mesoamerican city of Chichén Itzá—home to the best-preserved, biggest, and most elaborate playing court for what is often believed to be humanity's first team sport and one of the world's earliest ball sports. The sport known simply as the "Ball Game" was popular across Mesoamerica and played by all its major civilizations from the Olmec to the Maya to the Aztecs. It has been played since at least 1650 BCE and possibly as early as 2500 BCE.

Impressive stone ball courts were a staple feature of precolonial Mesoamerican cities, with many cities having multiple courts.

The simplistic earthen ball court found at the archaeological site of Paso de la Amada, a ruined city in what is now southern Mexico, is the oldest known surviving ball court, dating to 1400 BCE. But the stone ball court built around the year 900 CE at Chichén Itzá is the largest and most ornate playing field in Mesoamerica, representing the apogee of the Ball Game in the region.

Sports of one kind or another have been a part of most cultures past and present. Although some animals are known to play games with a physical aspect (e.g., pods of dolphins are known to play games of "catch" by tossing pufferfish back and forth like a ball), only humans have developed true sports—with rules and scores.

Sports are among humanity's oldest innovations. The earliest athletic competitions are believed to have been simple wrestling contests that are depicted in cave paintings. Other popular ancient sports included foot races, chariot races, and boxing, as well as weightlifting, swimming, and archery competitions. The Mesoamerican Ball Game was probably the first sport to contain the basic features of most modern team ball sports. (A rival for the title of earliest team ball sport is *cuju*, which some scholars believe may be older. *Cuju* is a Chinese game similar to soccer.)

Today, Chichén Itzá is a sprawling ruined city in the northern part of the Yucatán Peninsula in modern Mexico. Several prominent stone structures of the city remain well-preserved. Those include the Temple of the Warriors flanked by 200 columns carved in low relief to depict warriors, the Temple of Kukulcán (often called El Castillo), and the circular observatory known as El Caracol (the snail), named for the spiral staircase inside the tower. With more than 1 million visitors annually,

Chichén Itzá is among the most popular tourist destinations in Mexico. It is also a UNESCO World Heritage site, as well as an active archaeological site.

Chichén Itzá was once one of the greatest Mayan centers of the Yucatán Peninsula. The Mayan civilization was a Mesoamerican civilization noted for creating the most highly developed writing system in the Americas before the arrival of Columbus. The Maya are also famous for devising a sophisticated calendar and constructing monumental architecture, including pyramids.

Chichén Itzá was founded by the Itzá, a Mayan tribe or ethnic group. It was built near two earthen cavities forming natural wells or springs, which helped the people access underground stores of water in the area. The Yucatán Peninsula is a limestone plain that lacks rivers or streams but is marked with natural sinkholes or wells called *cenotes*. The city's name means "by the mouth of the well of the Itzá." Or including the literal translation of *Itzá*, "by the mouth of the well of the water-sorcerers." The construction of Chichén Itzá likely began in the fifth century.

The city rose to become a significant center of political, ceremonial, and economic activity in Mayan civilization by roughly 750 CE. By that time, Chichén Itzá had grown to become one of the largest Mayan cities, comprising nearly two square miles of densely packed stone buildings. A network of nearly 100 "sacbeob," or raised paved causeways, linked the city's structures. Those roadways and sidewalks were likely originally coated with limestone stucco or plaster, lending them a white color. Smaller structures sprang up on the outskirts of the city, composing Chichén Itzá's suburbs.

By the ninth century, Chichén Itzá was the de facto regional capital, with the city's rulers reigning over much of the northern and central Yucatán Peninsula. Chichén Itzá exhibits remarkable architectural similarities with the Toltec city of Tula, located nearly 1,000 miles away, including nearly identical temples. Attempts to explain the link between the two cultures have led to some controversy.

Some scholars believe Toltec warriors conquered Chichén Itzá in the 10th century; some believe the Toltecs influenced Chichén Itzá via cultural diffusion thanks to frequent trade exchange. Other theories abound. In any case, Chichén Itzá became a mixing bowl of the Toltec and Mayan cultures. The city of Chichén Itzá's most prominent temple is dedicated to the Toltec deity Quetzalcoatl, whom the local Maya adopted and called Kukulkán.

If you could visit Chichén Itzá in its heyday, you would have been struck by its colors. Today, the city's stone remains have faded to various shades of gray; however, archaeologists believe that originally the city's buildings were brightly painted. You would have passed by structures in intense shades of red, green, and blue, including the azure pigment now known as "Maya blue."

The wealthy wore similarly colorful dyed clothes made from animal skins, as well as elaborate feathered headdresses and ornate jewelry, such as beaded necklaces made of gold, turquoise, and jade. Men wore more jewelry than women, and taller headpieces often indicated higher status. Among those of lower status, men wore simple kilt-like wraps or loincloths, secured with a knot or woven belt, whereas women wore tunic-like *huipil* blouses and long skirts. Both men and women wore shawls wrapped around their shoulders during cold weather. Evidence indicates that some Maya anointed themselves with perfumes, made from vanilla or from flowers.

Surviving chemical residues suggest the ancient Maya of the Yucatán Peninsula traded food in open-air marketplaces. In the marketplace of Chichén Itzá, you would have likely seen a rich variety of foods, including avocados, plantains, limes, sour oranges, habaneros, chaya, cacao (chocolate), achiote, and fish, as well as meat from creatures ranging from deer to armadillos. The staple of the Mayan diet, maize, would have been ever-present. It was often boiled in water with lime and consumed as a gruel or porridge mixed with chili pepper, or was made into a dough for baking into tortillas, flat cakes, or tamales.

The Maya are believed to have had no currency and to have used a barter system instead. Via trade you may have been able to purchase goods ranging from ceramic pottery to woven blankets. The port at Isla Cerritos on Chichén Itzá's northern coast made the city an important commercial center, facilitating trade with other cities throughout the Americas. The people imported various goods from far away, such as red cinnabar pigment from the remote Guatemalan highlands.

At its peak, Chichén Itzá had as many as 50,000 people living in the city. That's similar to the population of Danville, Illinois, today, but it was the most populous city on the Yucatán Peninsula at the time. Its population was perhaps the most diverse in the Mayan civilization, with residents from across the Yucatán, Toltec migrants, and others originating from present-day Central America. The diversity may have stemmed in part from its status as a commercial center conducting frequent trade with distant peoples.

In the northwestern part of the city, you might have passed the *tzompantli* or ceremonial wall of skulls from victims of human sacrifice—despite its contributions to athletics, Chichén Itzá is not a place where a modern person would wish to live. In the distance, you might have heard the faint roar of cheering sports fans. If you kept walking, you would have come upon the great ball court and witnessed a game of the first team ball sport.

The great ball court of Chichén Itzá stretches a massive 225 feet wide and 545 feet long. The ancient sports arena's stone platforms measure 95 feet long and 25 feet high. At either end of the court, on the stone walls about 20 feet above the ground, jut out stone hoops. The hoop rings are engraved with intertwined feathered serpents—depictions of the deity Kukulkán.

The court has spectacular acoustic qualities. The temples lining the ends of the court contribute to a strong echo, so that something said at one end of the court can be heard on the other side and throughout the breadth of the court. This remarkable sound transmission helped make Chichén Itzá the preeminent ball court of the Mayan civilization, amplifying the cheers of fans and the calls of the ballplayers into a deafening din.

The sides of the court are lined with benches for onlookers. These benches are sloped to help keep the ball within the court. The benches are also carved with intricate reliefs showing past Ball Game victors holding aloft the decapitated heads of their opponents. Successful ballplayers were treated as celebrities in Mayan society, showered with riches and acclaim.

Some carved panels portray teams of 11 players, plus a team captain, whereas others show teams of 12 and a captain, suggesting some level of variation in the game rules. The precise rules of the game are unknown, but it is believed players passed a rubber ball across the field and knocked it through the stone hoops to score points.

At the end of many games, the losing team was beheaded and sacrificed to the Mayan deities. As Harvard University psychologist Steven Pinker has noted, the extent of human sacrifice in ancient Mesoamerica serves as a stark reminder of the ubiquity of violence in the past and of how far humanity has come since then.

That said, the Ball Game occasionally served as a substitute for war, with rival political leaders in the later Aztec civilization purportedly agreeing to confront each other on a ball court rather than on a battlefield. In fact, some psychologists believe that sports today help humans channel their competitive and aggressive impulses away from violence, and that athletic competitions are intertwined with the decline of overt conflict between states.

Chichén Itzá's population began to decline by the mid-13th century, when the seat of regional power within the Mayan world shifted to Mayapan, a newer city built to the southwest of Chichén Itzá. In the 16th century, Spanish conquistadors constructed a temporary capital there, before eventually abandoning it.

Today, Chichén Itzá is best known as one of the most famous and frequently visited of Mexico's Mayan historic sites. The city was voted one of the New Seven Wonders of the World in a global survey and attracts tourists from across the globe to marvel at its architecture.

A direct descendant of the Ball Game, *ulama*, is still played today—thankfully minus the ritual killing of the losing team. As such, the Ball Game is also the oldest continuously played ball sport in the world.

The development of team sports was a significant cultural achievement. Sports have transformed the way people spend their leisure time by being one of the most universally loved forms of entertainment. To many people, team sports fulfill deeper psychological functions, such as providing an additional sense of meaning in their lives.

Team sports enrich humanity because they are an exciting, aesthetically pleasing venue for emotional expression, an outlet for physical energy, an escape from real-world troubles, or a substitute for real-world conflict. For those reasons, Chichén Itzá has scored a place in history as a center of progress.

10
Athens
PHILOSOPHY

Our next center of progress is Athens during the Classical era (the fifth and fourth centuries BCE) in general and in particular the golden age of peace and cultural flourishing between the end of the Persian Wars and the start of the Peloponnesian War, spanning from 449 to 431 BCE. The city-state of Athens greatly valued intellectual pursuits and open inquiry, leading to the development of "philosophy," meaning the love of wisdom. Athenian philosophy encompassed natural philosophy (attempts to understand the natural world), as well as moral philosophy (ethics), metaphysics (theories about the fundamental nature of existence), and political philosophy. Athens was also the world's first (if restricted) democracy and has been nicknamed the Cradle of Western Civilization.

Of course, many people made important contributions to philosophical thought long before the golden age of Athens. The writings of Confucius and Lao-Tzu in the sixth century BCE, for example, remain influential to this day, with the latter's work carrying an anti-authoritarian message that has resonated across generations. Still, the Classical Athenians' impact on intellectual history is noteworthy.

Although the concept of empirical research took off only after the Scientific Revolution began at the end of the Renaissance, the ancient Athenians' devotion to understanding themselves and the world around them represented a significant intellectual breakthrough in human history. By recognizing the importance of debate and truth seeking, Classical Athens inspired countless thinkers over the succeeding millennia and heavily influenced the world we all live in now.

Today, Athens is the capital of and the largest city in Greece. It is also one of the biggest economic centers in southeastern Europe. Moreover, it is home to Piraeus—one of the world's largest passenger ports. More than 600,000 people live in the city proper, whereas the broader Athens metropolitan area houses about 3.75 million inhabitants. Athens is a major tourism center, due to its many well-preserved historic sites. The city has been called "the historical capital of Europe." Athens contains two different UNESCO World Heritage sites: the Acropolis of Athens and the medieval Byzantine Monastery of Daphni.

Athens likely gets its name from the Olympian goddess of wisdom, Athena, who was also the city's patron deity. Some scholars think it is the other way around and the goddess gets her name from the city. Depicted as a beautiful, but stern-faced, maiden clad in either a flowing *chiton* or full armor, Athena was also paradoxically the goddess of both war and peace, as well as the goddess of craftsmanship and weaving. The famous Parthenon temple on the Acropolis was built to honor her and to serve as the city's treasury, with construction starting in 447 BCE and decoration of the structure continuing until circa 432 BCE. Another large temple to Athena, built circa 420 BCE in the Ionic architectural style, also stands in a prominent position on the Acropolis.

The Acropolis is among the most distinctive features of today's Athens, surviving all the way from the fifth century BCE. It is a cluster of buildings on a rocky outcrop overlooking the city. If you could visit Athens during the fifth century BCE, you would have been struck not only by its majestic architecture, but also by the city's liveliness and energy. The beating heart of Athens was its marketplace, or *Agora*, meaning "place where people gather." In the bustling and noisy Agora, located to the northwest of the Acropolis, you would have witnessed people exchanging not only goods and services, but also ideas.

The structures surrounding the market stalls of the Agora included a series of stone benches, various altars and temples (notably the Temple of Hephaestus), a building called the *Aiakeion* (named for a judge of the underworld in Greek mythology) where laws and legal decisions were displayed, and various stoas, or covered porticoes. The Royal Stoa displayed the full legal code of the city, whereas the Painted Stoa (so-called because it was covered in artworks) served as a gathering place to watch jugglers, sword-swallowers, fire-eaters, and other entertainers—but also orators and philosophers, who drew large crowds. (The Stoic school of philosophy draws its name from the structure.)

Among the permanent and temporary stalls of the Agora you would have seen goods for sale that included food, wine, oils, furniture, clothes, leather sandals, perfume, and products from faraway lands brought in through Athens's well-situated port. You would have been able to buy wood from Italy, grain and linen from Egypt, dates from Phoenicia, ivory from North Africa, and spices from Syria. During the fifth century BCE, Athens was an unusually open society, far more open than other Greek city-states. It was arguably the world's first global city. As best-selling author Eric Weiner put it in his book *The Geography of Genius*: "This openness made Athens Athens. Openness to foreign goods, odd people, strange ideas."

Athens embraced relatively free and wide-ranging trade, exchange of ideas, and the incorporation of foreign-born people into its society. Free foreigners living in Athens, called metics (the rough equivalent of resident aliens), enjoyed considerable social mobility and could attain high-status roles. The Athenians borrowed many of their ideas from abroad, importing the Phoenician alphabet, Egyptian medicine and sculpture techniques, Babylonian mathematics, and Sumerian literature. The Athenians often improved on what they borrowed. For example, although the Egyptians invented the art of sculpting statues, the Greeks were the first to carve truly realistic, lifelike human forms from stone. The philosopher Plato summed it up when he said, "What the Greeks borrow from foreigners, they perfect."

However, many of the foreign-born people in Athens were not free. Slavery was ubiquitous throughout the ancient world, and Athens was no exception. In the Agora, you would have been horrified to see human beings offered up for sale. Nearly all of the enslaved people in Athens were not of Greek origin, but were what the Greeks called *barbaroi* (barbarians) from abroad, often captured in conflicts from farther north. Many slaves thus had fair pigmentation that distinguished them from the native Athenian population, who tended to have dark hair and olive skin. Names like Xanthias (meaning "blond") and Pyrrhias (meaning "redhead") became virtual synonyms for "slave."

Most Athenian slaves suffered greatly, although the institution was relatively fluid compared with other Greek city-states at the time. Other Greeks were sometimes shocked at the blurring of boundaries between enslaved and free persons in Athens, or what the author now known as Pseudo-Xenophon called the "uncontrolled wantonness" of Athens's "equality between slaves and free men, and between metics and citizens." He was horrified to find that in Athens, some people legally classified as slaves accumulated great wealth, whereas some free people were terribly poor, thus obfuscating the distinction between slaves and free persons.

Walking away from the Agora, you would have come upon the limestone Bouleuterion, or meeting place of the Boule. The Boule was composed of 500 Athenian citizens, chosen by lottery to serve for a yearlong term. They met in the assembly building every day (with the exception of festival days) to prepare legislation for the review of the *ekklesia*, or assembly of all voting citizens.

Often called the first democracy, Athens allowed each of its adult male citizens a vote when deciding on official policies. In the mid-fifth century BCE, the number of eligible voters was perhaps as high as 60,000. (That population would decrease significantly during the Peloponnesian War, when many Athenian men perished.) Still, that was no more than 10 to 20 percent of the city's population, with the majority of Athenians being excluded from political participation because of their sex or citizenship status. Women, metics, and slaves were given no vote. Although the world's first democracy was deeply flawed, the Athenian experiment influenced the evolution of modern representative democracies.

In the fifth century BCE, Athens served as home to a great number of geniuses and innovators, including the playwrights Aeschylus (525/524–456/455 BCE), Sophocles (c. 496–406 BCE), and Euripides (c. 484–406 BCE); the historians Thucydides (c. 460–after 404 BCE) and Herodotus (c. 484–c. 425 BCE); and the hugely influential philosophers Socrates (c. 470–399 BCE) and Plato (428/427–348/347 BCE).

Socrates developed the "Socratic method" of inquiry, which uses questions to stimulate critical thinking, draw out ideas, and unveil assumptions. During the 18th-century Enlightenment, many thinkers, including Voltaire, drew inspiration from Socrates, who was upheld as an early advocate of reason. Socrates's student Plato became the father of the philosophical school of thought known as idealism and is often considered to be the founder of Western political philosophy.

Socrates's persistent and public questioning of prominent Athenians often left the latter looking foolish, making him many enemies. After Sparta and its allies defeated Athens in the Peloponnesian War, Athenian society entered a period of social and intellectual upheaval in which the city's commitment to free speech and open inquiry wavered. Perhaps in part looking for a scapegoat for the city's various misfortunes, accusers charged Socrates with "impiety" and "corruption of the youth." The philosopher was sentenced to death. Ironically, the same city that arguably served as the world capital of philosophy and critical thinking executed a man for the crime of simply asking questions.

Before the arrival of philosophy, society largely focused on immediate and practical concerns and did not devote significant amounts of time or effort to seeking out knowledge for its own sake. Philosophy represented a shift in priorities in intellectual life. By valuing wisdom as its own end, Athens encouraged people to devote their minds to contemplation of (and development of systematic theories about) morality, society, the workings of the universe, and so forth. Human beings are intrinsically curious, but Athens helped elevate curiosity to a moral imperative.

It also created institutions to support the inquisitive nature of mankind. By the fourth century BCE, Athens could also count the philosopher Aristotle (a student of Plato) among its luminaries and became home to the forerunners of modern universities. Those forerunners included Plato's Academy—the first true institution of higher learning in the Western world and a prototype for later universities—and Aristotle's Lyceum, a temple that served as a center for education, debate, and scholarship.

Aristotle (384–322 BCE) is often considered the founder of the field of formal logic, and he also made groundbreaking contributions to numerous disciplines, such as physics, chemistry, biology, history, literary theory, and, perhaps most significantly, political theory and ethics. He promoted what has come to be called "virtue ethics" as a practical means to *eudaimonia* or human flourishing, and believed that governments ultimately exist to foster virtue. (For instance, he defended private property on the grounds that communal property can lead to disagreements and disincentivize the virtue of generosity.)

Today, Athens is still best known for its far-reaching influence as an intellectual center of the ancient world. The image of Athens in the popular imagination is perhaps best summed up by the famous 16th-century fresco titled *The School of Athens*. Created by the Italian Renaissance artist Raphael to decorate the Apostolic Palace in the Vatican, the painting depicts many of the most influential Athenian philosophers of the Classical era engaged in passionate debate with one another, writing down their ideas, passing on their knowledge to pupils, or otherwise devoting themselves to the pursuit of truth.

Athens played a key historical role in promoting the importance of open inquiry, reason, debate, and the pursuit of truth. Athens created centers of scholarship that were the forerunners of modern university systems, established a new approach to understanding the natural world that acted as the precursor to modern science, and experimented with a new system of government that would one day inspire the creation of modern representative democracy. For heightening humanity's "love of wisdom," Athens is rightfully a center of progress.

11

Alexandria

INFORMATION

Our next center of progress is Alexandria during the third and second centuries BCE, when the Great Library marked the city as, arguably, the intellectual capital of the world. During the third century BCE, an educational and research institution called the Mouseion (literally, "shrine of the Muses"), from which we get the word museum, was built in Alexandria. The Great Library of Alexandria was one part of the Mouseion. Although estimates vary widely, the library may have held about 700,000 scrolls, the equivalent of more than 100,000 printed books. The amalgamation of so much written knowledge in one place represented a breakthrough in the way that humanity stored and distributed information.

For people today, who have grown up with unparalleled access to information thanks to the internet, it is difficult to comprehend a world where information is out of reach. But throughout much of history, knowledge often went unwritten. Even when written down, information was typically scattered in different places or otherwise inaccessible.

In the Great Library of Alexandria, much of humanity's collective knowledge of subjects ranging from medicine to astronomy could be accessed in a single place. Among the writings that you could browse in the library were histories, philosophical treatises, literary works of poetry and prose, and the *Pinakes*—believed to be the world's first library catalog. Philosophers and scholars flocked to the city, attracted by its library's vast compendium of information and the city's reputation as an intellectual center.

Alexandria was founded in 331 BCE by the Macedonian leader Alexander the Great (356–323 BCE), who was in the midst of conquering the Persian Empire. Alexander drove out the Persian invaders who had deposed the last indigenous king of ancient Egypt just over a decade prior. Alexander departed from Egypt a few months after founding Alexandria, leaving his viceroy Cleomenes in charge.

After Alexander passed away in 323 BCE, one of his deputies, a Macedonian general by the name of Ptolemy I Soter, took control of Egypt. Ptolemy executed Cleomenes and declared himself pharaoh. He started what came to be known as the Ptolemaic dynasty and made Alexandria his capital in 305 BCE. The Ptolemy family—despite a seemingly hereditary tendency toward morbid obesity and lethargy—managed to stay in power until 30 BCE.

The city's population rapidly grew to about 300,000 people. Alexandria became a key center of Hellenistic civilization. It remained the capital of Ptolemaic Egypt, as well as Roman and Byzantine Egypt, for almost a millennium (until the Muslim conquest of Egypt overseen by the Rashidun Caliphate in 641 CE). Alexandria was also the largest city in the ancient world, until Rome eventually grew even larger.

Today, Alexandria is the second-largest city in Egypt. It is a major economic center and the most populous city on the Mediterranean Sea. It has a population of more than 5 million people. Alexandria is thus also the sixth-largest city in the Arab world and the ninth-largest city in Africa. Its historical importance makes it a well-frequented tourist destination. It is also a major industrial center, due to its pipelines of natural gas and oil from the Suez.

If you were to sail to Alexandria during the time of its famed library, you would have been struck by the towering sight of one of the Seven Wonders of the Ancient World. Hellenistic Alexandria was home to one of the most impressive and famous sites in antiquity, the Pharos or the great lighthouse, which was constructed in the third century BCE. Standing at least 330 feet high (and possibly higher), the Pharos was taller than the Statue of Liberty (305 feet) and Rio de Janeiro's iconic Christ the Redeemer statue (125 feet). For many centuries, the Pharos remained among the tallest manmade structures in the world. At the top of the lighthouse's tower, a fire, which was likely kept burning with oil rather than wood, lit the way for ships entering Alexandria's harbor.

Sailing nearer, you would have seen the city of Alexandria emerge on an isthmus opposite the small island on which the

Pharos stood. You would have viewed the city's classical architecture laid out among the orderly parallel lines of the city's streets. Alexandria was designed by the architect Dinocrates of Rhodes, using a Hippodamian grid street plan. After docking at the harbor and setting foot in the city itself, you would have observed a wide variety of people, with the three most common ethnicities being Greeks, Jews, and Egyptians.

In other words, the city was cosmopolitan and diverse. In the city's southwest was the Rhakotis—a settlement predating Alexandria that had been absorbed into the city. It was mainly inhabited by the pre-Arab Egyptians, who may have been a mix of African, Semitic, and Hamitic ethnicities. Some of those residents may have continued to wear the Egyptian kilts, tunics, and dresses that had been common before Alexander the Great's conquest and Ptolemaic rule. However, many urban Egyptians adopted Hellenized clothing as a social marker of status. The Jewish quarter in the city's northeast was home to one of the largest urban Jewish communities in the world at the time. During the city's golden age, Alexandria was tolerant of religious differences. Notable Jewish Alexandrians included the historian Artapanus of Alexandria, Demetrius the Chronographer, and the playwright known as Ezekiel the Tragedian.

The Broucheion was the wealthy Greek or Royal quarter of Alexandria, which contained the city's grandest architecture. Most people there would have worn Greek garments, such as the *himation* and *chiton* or highly Hellenized versions of traditional Egyptian clothing. In the Broucheion, you would have seen magnificent temples to the Greeks' deities—prominently Poseidon, the god of the sea. Alexandria was, after all, a coastal city dependent on maritime trade. The Broucheion also contained a theater, and you could have seen theatergoers milling about it, discussing the latest plays. Alexandria had a thriving arts scene. The city was famous for its professional entertainers, which New York's Metropolitan Museum of Art calls "a combination of mime and dancer," as well as its poets and playwrights.

Within the royal palace's grounds in the Broucheion, you would have found the Mouseion and the Great Library—two beautifully decorated edifices on a campus of architecturally intricate buildings and flowering gardens. The Mouseion building included a lengthy roofed walkway and a large communal dining hall, where scholars dined and exchanged thoughts. The Mouseion also contained exhibit halls (from which we derive the modern notion of "museum"), private study rooms, lecture halls, residential quarters for scholars, and theaters for live performances. The Great Library consisted of shelves upon shelves of papyrus scrolls.

The Mouseion was most likely founded by the first Ptolemy king, Ptolemy I Soter, who is thought to have entrusted the creation of the Mouseion and the Great Library to Demetrius of Phaleron—a former Athenian politician, who had fallen from power in his home city-state and sought refuge at Ptolemy's court. A letter surviving from the second century BCE reveals that the new institution was envisioned as a universal library that would encompass all of the world's written knowledge:

> Demetrius . . . had at his disposal a large budget in order to collect, if possible, all the books in the world . . . [T]o the best of his ability, he carried out the king's objective.

The library soon compiled the whole corpus of Greek literature, including the books of Aristotle, along with various texts in other languages, such as Egyptian. The Mouseion's scholars produced many new works to add to the shelves.

The Mouseion was a research institution with over 100 scholars living and working in the complex at any given time. The Mouseion's scholars were salaried employees, who were motivated, in part, by monetary incentives. In addition to their salaries, for example, they received free room and board and paid no taxes. Their expertise spanned a range of disciplines. One room in the Mouseion was dedicated to the study of anatomy; another area was dedicated to astronomy; and so forth. A famous medical school was also established at the Mouseion, where Galen would study centuries later. His contributions to human medical understanding were numerous and varied. Papyrus scrolls in the Great Library likely chronicled everything from mental disorders to intestinal diseases, from surgery and bone setting to dentistry, and even the making of false teeth.

It was thanks to the Great Library that the scholars of the Mouseion were able to achieve so much. The library marked Alexandria as the world capital of information, attracting many of the brightest minds of the day. In Alexandria, the astronomer Aristarchus (c. 310–c. 230 BCE) theorized that the Earth revolves around the sun. He did so 1,800 years before Copernicus. The physician Herophilus (possibly 325–255 BCE) first identified the brain as the organ that controls the movement of the body. The Egyptian priest Manetho (early third century BCE) chronicled Egypt's pharaohs and organized Egyptian history into dynasties still used by historians today. The poet Callimachus (c. 305–c. 240 BCE) detailed the texts in the library, which were organized by subject and author, thus creating the first library catalog and becoming the father of library science.

The inventor and mathematician Archimedes (287–212 BCE) studied in Alexandria and may also have taught there. While taking a bath, Archimedes realized that displaced water could be used to measure the volume of an object. He is said to have shouted *Eureka!* (Greek for "I have found it!"), leapt from the bathtub without bothering to dress, and ran through the streets to announce his discovery. The geographer Eratosthenes (c. 276–194 BCE) also taught in Alexandria and made his groundbreaking calculation of the circumference of the Earth in that city. He was the founder of chronology, the first person to calculate the tilt of the Earth's axis, and the creator of the first global map projection of the world. (In cartography, a map projection is a precise method of displaying the globe's surface as a flat plane while maintaining accuracy.) The founder of the mathematical subdiscipline of geometry, Euclid (born c. 300 or c. 325 BCE) taught at Alexandria, too. Later, the engineer and mathematician Hero, also called Heron (possibly 10–70 CE), nicknamed the Greatest Experimenter of Antiquity, lived and worked in Alexandria as well. It was there that he invented the aeolipile—the first known device to transform steam into rotary motion. (At the time, the steam turbine was treated as an amusing curiosity without any practical purpose.)

The Great Library and the Mouseion were open to scholars from all cultures and backgrounds. Both women and men were allowed to study the texts in the Great Library. A few centuries after the period that concerns us, one of the first recorded female scholars, the philosopher and mathematician Hypatia (born between 350 and 370 CE and died 415 CE) would work in Alexandria, in the spirit of the Mouseion even after that institution's destruction.

Other cities had built libraries before; however, Alexandria pioneered the idea of a universal library at a scale never before achieved. Libraries and archives were kept in many cities in various ancient civilizations, including Egypt, Mesopotamia, Syria, and Greece. However, those earlier institutions were limited in scope, typically only containing local knowledge or covering a particular subject area, and were chiefly oriented toward conservation of a particular cultural tradition or heritage.

The idea of a universal library, like that of Alexandria, proved game-changing. Alexandria's library contained works concerning practices from far away. To give an example, it included scrolls describing Buddhism, which arrived in the library as a result of diplomatic exchange between India's Ashoka the Great and Ptolemy II Philadelphus. Alexandria inspired other cities to create rival "universal libraries," such as the Library of Pergamum, sometimes referred to as Pergamon, in what is today Turkey.

The Great Library was eventually destroyed. The main library structure was likely burned in 48 BCE, when the last Ptolemy ruler, Ptolemy XIII, laid siege to his wife, sister, and coruler Cleopatra and her lover, the Roman dictator Julius Caesar. The secondary library building—which was in the Serapeion temple and was added when the first library could fit no more scrolls—may have survived until near the end of the fourth century, when the Byzantine emperor Theodosius I ordered the demolition of all pagan temples.

For seeking to compile all of the known knowledge in the world in one place and make it accessible to scholars from all parts of the Mediterranean, Alexandria during the third and second centuries BCE and beyond deserves recognition as a center of progress. Alexandria pioneered the concept of a universal library. Long after the Great Library of Alexandria ceased operating, people have continued to expand the store of human knowledge, and access to it, ultimately culminating in tools like internet search engines and crowd-sourced encyclopedias or wikis. Today, many of us carry the keys to a library that is infinitely larger than Alexandria's—in our pockets in the form of smartphones.

12

Rome

R O A D S

Our next center of progress is ancient Rome during its Republican and early Imperial periods, when the Romans built infrastructure projects that were, at the time, unparalleled in their sophistication. Those projects ranged from aqueducts and sewers to bridges, amphitheaters, and bathhouses. The *viae Romanae* (Roman ways) or Roman road network, in particular, represented a breakthrough. Built in part to ease the transportation of soldiers and the delivery of military supplies, the roads also greatly aided the free movement of civilians and trade goods. The Romans pioneered such new concepts as milestone markers, advanced surveying, and various engineering marvels, such as viaducts, to generate the shortest and straightest possible routes.

Although the Romans did not invent roads—a Bronze Age innovation—the Romans vastly improved on roads' concept and potential. As early as 4000 BCE, the older Indus valley civilization created paved and straight roads intersecting one another at right angles. But the sheer scale of the later Roman road network and the institution of several important innovations would forever alter the way people travel.

Today, we take advanced road systems for granted, but reliable roads were once a rarity, and many journeys, of course, took place with no roads at all. By making travel faster and easier, Roman roads greatly increased the efficiency of transporting trade goods, people, and messages. The Roman road system increased the rate of cultural exchange and encouraged connections that helped unify the Roman Empire—a melting pot of different cultures, beliefs, and institutions.

Major roadways within the Roman road system were typically paved with stone and flanked by bridle ways or horse trails and footpaths to separate different kinds of traffic. The roads were also often cambered to allow rainwater to drain into parallel ditches or gutters. At the peak of Rome's strength and influence, the empire's provinces were interconnected by perhaps 372 great roads, and an estimated 29 major highways radiated from the city of Rome itself. Thus, the popular expression, "All roads lead to Rome."

Today, Rome is the capital of Italy and the country's most popular city for tourism. It is also a major European business center and the seat of several United Nations agencies. In addition, it is home to the pope, also known as the bishop of Rome, who is the head of the Catholic Church and resides in the independent state of Vatican City, which was carved out of Rome in 1929. Located in the central-western portion of the Italian Peninsula, Rome is among the oldest continuously occupied cities in Europe. Many historians consider it to be the world's first-ever imperial city and true metropolis, although others might point to alternatives such as Pataliputra in the Mauryan Empire as holding that title. Rome's nicknames include the Eternal City (*Urbs Aeterna* in Latin; *La Città Eterna* in modern Italian) and *Caput Mundi* (Latin for the Capital of the World).

Tradition holds that Rome was founded in 753 BCE, although the area was inhabited beforehand. According to legend, the sister of the ruler of Alba Longa, a Latin city in Central Italy, gave birth to twin boys who were ostensibly fathered by Mars, the Roman god of war. The ruler saw the newborns as a threat to his rule and forced his sister to abandon them. The twin brothers, Romulus and Remus, were then allegedly nursed by a female wolf and adopted by a shepherd. They grew up to lead a successful rebellion against their uncle and reinstate their grandfather as king. After doing so, they returned to the hills (i.e., the famed Seven Hills of Rome), where they decided to build a city. A disagreement about the precise location of the city (Romulus is said to have preferred the Palatine Hill and Remus is said to have preferred the Aventine Hill) led Romulus to kill his brother. After founding Rome, Romulus reigned over it as its first king.

Ancient Roman history is typically divided into three eras based on the city's evolving governing structure: the Period of Kings (625–510 BCE), Republican Rome (510–31 BCE), and

Imperial Rome (31 BCE–476 CE). In keeping with the myth of being founded by a son of the god of war, throughout much of its history, Rome was in a state of conflict. It served as the capital of a polity that often sought to expand its territory. At its peak, the Roman Empire encompassed an area of about 2 million square miles. It contained modern-day Spain, Portugal, France, Belgium, parts of Germany, England, Wales, much of central and southeastern Europe, Turkey, parts of Syria, and lengthy territory along the coast of North Africa, including a substantial portion of modern Egypt.

To support their expansionism, the Romans eventually formed a large, elite professional army. Motivated in part by the need to move their soldiers across vast distances, the Romans created their extensive road network, the remnants of which are still visible across much of Europe and parts of North Africa and the Middle East. Not until the Inca Empire's road network, 1,000 years later, would a comparably complex road system arise. (The paved portion of the Roman network was twice as many miles long as the Incas' road system.)

The first major road constructed by the Romans was the Appian Way, which connected the city of Rome with Capua, on the northeastern edge of the Campanian Plain. Construction of the Appian Way began in 312 BCE, during the Republican period when Rome was governed by an unelected senate and officials called consuls. (It should be noted that the city's republican system was oligarchical—with a few wealthy families maintaining most of the power—and not a democracy.)

By circa 244 BCE, the road was extended to stretch past Capua to Brindisi, a port city on the Adriatic Sea, located in southeastern Italy's Apulia region. The Appian Way's praises were sung by the poets Horace and Statius, who called it *longarum regina viarum* (queen of long-distance roads). As the best route to the seaports of southeastern Italy, and thus an important gateway to Greece and the eastern Mediterranean, the Appian Way was of tremendous strategic importance.

Although the Appian Way was first built to speed the delivery of military supplies during the Samnite Wars, it proved to be the first in a series of highways with an importance that went far beyond military uses. If you could visit Rome a few centuries later, during the era of Augustus Caesar (63 BCE–19 CE)—the grandnephew and adopted son of Julius Caesar, who made an appearance in our previous chapter—when the road network was already well established, you would have entered the thriving capital city of a far-reaching empire connected by the *viae Romanae*.

Augustus took advantage of growing cynicism toward the republic, which came to be widely viewed as corrupt, to seize absolute power. He pretended that he was not a dictator, taking the title First Citizen instead. Augustus bought the public's support through the expansion of the Roman welfare system—which would eventually reach unsustainable levels. He also instituted a series of (to the modern eye) bizarre, sexist, and draconian morality laws known as the Augustan *Leges Juliae*, which legalized the murder of alleged adulterers in many cases if the accused was female and pressured widows to remarry. The laws were poorly received and short-lived.

However, Augustus's reign also saw the beginning of an era of relative peace known as the *Pax Romana*, in which Rome avoided entanglement in a major war for almost two centuries, although it continued small-scale wars of expansion. Benefiting from this relative peace, as well as an exceptional trade network aided by the Roman Empire's roadways, the city of Rome grew and prospered. As a visitor, you would have been mesmerized by the imposing architecture of its massive buildings and bustling crowds of diverse people moving through its streets, typically clad in tunics. Men wore knee-length *chitons* and, sometimes, togas. Women wore ankle-length tunics and, sometimes, woolen *stolae*, like the one tied at the shoulder of the Statue of Liberty.

Like all ancient civilizations, Romans practiced slavery, and many people, even in skilled positions, such as accountants and physicians, were enslaved. A modern person would also be horrified by the degree of poverty in the city. But for its era, Rome was among the wealthiest places on Earth. The city of Rome itself held about 1 million inhabitants at the time or was at least fast approaching that number. That constituted an urban population not equaled again in any European city until the 19th century. Although about the same size as the population of today's San Jose, California, it was then a metropolis of unrivaled magnitude.

At the center of the city was the Roman Forum, a rectangular travertine-paved plaza flanked by several significant buildings. Romans referred to this space as the *Forum Magnum*, or simply as "the Forum." Originally the city's marketplace, the Forum became the city's civic center during the Republican era. It was home to public meetings, law court sessions, and gladiatorial fights, and remained lined with shops that formed an open-air marketplace. In the period that concerns us, circa 20 BCE, the Forum's primary role was beginning to shift to serve as a center for religious and secular spectacles and ceremonies. It was also the endpoint of celebratory military parades or processions known as triumphs.

Entering the Forum in 20 BCE, you might have witnessed the erection of the *Milliarium Aureum* (Golden Milestone). The Golden Milestone was an important monument, likely measuring about 12 feet high and constructed of marble that was sheathed in gilded bronze. It stood near the prominent Temple of Saturn in the bustling central Forum. The monument was the symbolic and practical midpoint of the Roman road system. All roads were considered to begin at the Golden Milestone, and all distances in the Roman Empire were measured relative to the monument. To this day, a marble structure thought by some to be the base of the monument can be viewed in Rome.

The monument's dedication ceremony would have been an exciting affair, perhaps involving festivities, lofty speeches, and a large crowd. The Golden Milestone represented the achievement of connecting much of the world through a network of reliable roads—enabling travel, transportation of goods, and faster delivery of messages.

Most roads were winding and uneven and built to accommodate natural obstacles; however, the Romans prided themselves on creating straight roads. Instead of having their roads wind around natural obstacles, Roman engineers found ways to continue straight ahead by building bridges, tunnels, or viaducts. They would also drain marshes, cut through forests, or divert the paths of creeks when needed.

Before a road was built, extensive surveying was carried out to find the shortest and straightest possible route between two points and to determine what engineering feats would be needed to tackle any obstacles in the way. A surveyor ensured that the land was level and a suggested path marked out with wooden stakes. He would have used a tool called a *groma* (a wooden cross with weights) to make certain that the lines were straight. Once the path was decided on, the Romans would create earthen banks called *aggers* on which they would lay the road material, and dig a ditch on either side for drainage.

Roads were sometimes built in several layers, with stone blocks overlying crushed stone or gravel in cement atop stone slabs (also in cement) above crushed rock over a base layer of compacted sand or dry earth. These layers gave Roman roads their longevity. While other roads quickly wore down into sunken muddy trails, Roman roads lasted for centuries or even millennia. The Romans also instituted a system of regular milestone markers and standardized road widths. Moreover, they experimented with grooved roads to assist with the transport of wheeled carts and chariots.

Rome remains best known for its historical influence, including its far-reaching empire and its fervent rejection of monarchy during the Republican era. The latter would later help inspire the U.S. Founding Fathers. Roman infrastructure projects from the days of the empire left a permanent mark on the world that's rather wryly summed up by a scene in a classic British comedy film, *Monty Python's Life of Brian*, in which a gathering of people plotting a rebellion against the Romans nonetheless concede that the Romans created great aqueducts, roads, and so forth.

Roman baths are still in use in Algeria, two millennia after being constructed, and a Roman amphitheater in France, the Arena of Nîmes, still holds live concerts today. In Rome itself, a section of the *Cloaca Maxima* (Greatest Sewer), dating to the Augustan period, is still in use. But it was Roman roads that arguably left the greatest mark of all. To this day, many of the roads survive, and some of their alignments are still in use—with modern roads overlaying the original routes. For example, parts of Great Britain's road system run along old Roman routes—such as much of the 18-mile section of the A1 road that links Dishforth and Catterick. Although it is no longer true that "all roads lead to Rome," as the saying goes, many do.

For taking the concept of a road to new heights, creating the greatest road network of the ancient world, and proving the possibility of such a comprehensive, efficient, and lasting road system, Rome is rightfully a center of progress. Numerous Roman paths, in areas ranging from western Europe to North Africa, are still traveled today. Rome showed the world the potential of roads to increase the efficiency of travel, the transportation of goods, and the delivery of information.

13

Chang'an

TRADE

Our next center of progress is Tang dynasty–era Chang'an, the easternmost stop along the Silk Road, once the world's longest trade route. Many historians regard the Tang dynasty (618–907 CE) as a high point in Chinese civilization—a golden age of cosmopolitan culture. The Tang dynasty capital Chang'an was among the most prosperous and populous cities in the world, with over 1 million inhabitants within the city walls (and perhaps 3 million including the city's suburbs) by the end of the dynasty. Although many places featured as centers of progress owed their prosperity at least in part to robust trade, perhaps no city in the ancient world better epitomized the enriching effects of trade than Chang'an.

The Silk Road linked many civilizations, such as the Roman Empire and the Chinese Empire. Great caravans of merchants traveling along the Silk Road carried out early trade in fabrics like silk and wool, precious metals like gold and silver, and other goods. The Silk Road stretched 4,000 miles, and took international trade and cultural exchange to new heights, connecting East and West. It was on the Silk Road that the East and West exchanged not only goods but also ideas. As such, the Silk Road was also the world's preeminent long-distance communication network.

Chang'an stood in the central area of the present-day city of Xi'an (Western Peace), the capital of Shaanxi Province. With more than 12 million people, Xi'an is the most populous city in northwestern China and has been called an emerging Chinese megacity or megalopolis. The urban center remains famous for being one of the oldest cities in China and the Silk Road's so-called starting point. It is also the home of the famed "terra-cotta army," a collection of thousands of statues of soldiers buried with China's first emperor to guard him in the afterlife.

Chang'an (Perpetual Peace) was an ancient capital of more than 10 dynasties in Chinese history. Its name is appropriate, given its significant role in international trade history because peace is a requirement for two countries to engage in trade— the nonviolent exchange of goods. The first emperor of unified China, Qin Shi Huang, had his mausoleum, brimming with terra-cotta soldiers, built between 246 and 208 BCE, slightly to the east of where Chang'an would appear. The traditional date given for Chang'an's founding is 202 BCE, at the start of the Han dynasty, when that dynasty's founding emperor chose to place his capital there. He had a palace built in Chang'an that was then the largest palace ever constructed on Earth, spanning 1,200 acres. It was fittingly named Weiyang (Endless Palace), and it survived until late in the Tang dynasty.

A complex network of trade routes emanating from Chang'an and extending to the heartland of central Asia first began to emerge between the second century BCE and first century CE. The Silk Road peaked between 500 and 800 CE, enabling large-scale, long-distance trade of valuable goods. The great caravan tract followed the Great Wall of China's path to the northwest, bypassed the Taklamakan Desert, crossed the Pamir Mountains in Tajikistan, traversed Afghanistan, and continued to the Levant, where merchandise traveled by ship across the Mediterranean Sea. Very few merchants made the journey across the entire Silk Road. Instead, goods were passed along in a staggered progression, with most merchants acting as go-betweens who only traveled along a Silk Road subsection.

The Silk Road takes its name from one of the most prized goods sold along its route. Each spring, the city of Chang'an held an imperial ceremony of silk production. The women of the court prepared fine silk, stretching and ironing the newly woven fabric to perfection. There exists a famous painting of Tang dynasty–era court women preparing silk during the ceremony. (One can view a replica of that painting dating to the later Song dynasty in the Boston Museum of Fine Arts.) Although Chang'an was famous for its silk exports, the Silk Road was a conduit of far more than silk. Chinese exports also included paper, rice wine, perfumes, camphor, and medicinal drugs. But imports are what made life in Chang'an so vibrant.

If you could visit Chang'an in the heyday of the Silk Road, you would enter a flourishing cosmopolis featuring the best of many different cultures and enveloped in a festival-like atmosphere. In Chang'an's streets, parades of performers acted out plays from Sogdiana, an ancient Iranian civilization, before cheering crowds. The streets featured all manner of traveling performances, including magic shows from as far away as Rome. Dancers from many locales, including a renowned troupe from Tashkent in today's Uzbekistan, performed in Chang'an's boisterous taverns.

Amid the city's elaborate architecture, you would have seen beautiful temples and the soaring Dayan Pagoda (still standing) that housed a Buddhist library featuring Indian scriptures. The teeming marketplace (now a museum) overflowed with novel goods brought by caravans of foreign merchants, including carpets from Persia, ivory from what is now Thailand, spices from India, and Roman glassware. Around the city, you would have encountered diverse peoples and heard many different languages.

Enriched by trade, Chang'an flourished and became the site of a series of beautiful palaces built by the emperor to showcase his empire's prosperity. The emperor's court was known for its many hundreds of dancers, and the court also maintained at least nine different musical ensembles. Each ensemble specialized in a different musical style stemming from various lands. The musicians used imported instruments, such as cymbals from India and lacquered drums from Kucha—an ancient Buddhist kingdom along the Silk Road. According to the *Encyclopedia Britannica*, "One can sense in Tang musical culture an internationalism not matched until the mid-20th century, when radios and phonographs provided their owners with the delights of a similarly diverse and extensive range of choices."

Not everyone appreciated the fruits of cultural exchange. The eighth-century poet and government official Yuan Zhen, who considered non-Chinese persons to be "barbarians," lamented the presence of foreign people and practices in China. He complained of alleged air pollution created by foreigners, decried Chinese women who wore imported makeup, and complained of entertainers who devoted themselves to foreign musical styles. Although he wrote those words in connection with his home city of Luoyang, the effects of cultural exchange would have been even more pronounced in Chang'an:

> Ever since . . . the Western barbarians kicked up smoke and dust, the rank stink of fleece, felt, and mutton has pervaded Luoyang. Our women have become barbarians' wives and learned to apply their makeup, while singing girls offer barbarian tunes and focus on barbarian music.

(Note that to Yuan Zhen, "Western" meant anything beyond the Great Wall.)

The Silk Road not only enriched Chang'an's artistic scene, but it also introduced many new ideas to the city. Various philosophies and religions came to China along the Silk Road, notably Buddhism from India. The people of Chang'an also became familiar with Nestorian Christianity from Syria, Zoroastrianism and Manichaeism from Persia, Judaism, and Islam spread by Arab merchants. Muslims built the city's Great Mosque in 742. For a time, diversity of thought blossomed and the city was known for its tolerance of religious and philosophical differences.

However, as the Tang dynasty began to decline, xenophobia and religious intolerance grew. Moreover, as Chang'an grew wealthier, it unfortunately became a target for military attacks and the city became unstable. It was captured by rebel forces led by a general named An Lushan, in 756, but was retaken by the Tang dynasty the next year. In 763, invaders from the

Tibetan Empire briefly occupied Chang'an and an alliance of the Tibetan Empire and the Uyghur Khaganate again besieged the city in 765. Tensions gave rise to two notable massacres of Silk Road merchants, led by anti-Tang rebel army leaders— Tian Shengong and Huang Chao, respectively. The first was the Yangzhou massacre (760), followed by the Guangzhou massacre (878 or 879). Both massacres involved the slaughter of more than 100,000 Arab and Persian merchants. Among the victims were Muslims, Jews, Christians, and Zoroastrians.

A series of rebellions, including the one led by Huang Chao, ultimately proved devastating to the Tang dynasty. Huang Chao sacked Chang'an in 881. Although Tang forces were eventually able to suppress that rebellion and reclaim the city, the dynasty never fully recovered and was soon ousted. Further political instability in other parts of the world caused by the loss of several Roman territories in Asia and the rise of Arab power in the Levant made the Silk Road increasingly unsafe. Hence, trade along the route declined precipitously.

However, in the 13th and 14th centuries, the Mongol Empire brought the Silk Road back into common use. It was then that the writer and merchant Marco Polo (1254–1324) made his famous journey from Venice to China. The Silk Road embodied not only the potential of trade to improve lives and create prosperity, but also the challenges that come with global interconnectivity, such as the potential for cultural clashes and the spread of contagious illnesses. In the mid-14th century, the Silk Road helped spread the bacteria responsible for the Black Death pandemic from Asia to Europe.

Perhaps no city was more emblematic of the Silk Road than Chang'an. The city is often called the "starting point" of the

Silk Road because of its status as the easternmost stop along that trade route and as the origin point of much of that route's namesake silk. Trade brought Chang'an extraordinary cultural and economic wealth and made it one of the world's most dazzling and cosmopolitan cities in its day. For its vital connection to the Silk Road, which greatly expanded the international exchange of goods and ideas, Tang dynasty–era Chang'an is deservedly a center of progress. Today, global trade and cultural exchange have risen to heights that the Silk Road merchants could not have imagined. Although challenges such as pandemics remain a part of globalization, trade and exchange continue to enrich our lives immeasurably.

14

Baghdad

ASTRONOMY

Our next center of progress is ninth-century Baghdad, during the Abbasid Caliphate at the beginning of the so-called Islamic Golden Age. Baghdad was quickly growing into the world's largest city and was a major learning center that saw breakthroughs in mathematics and, most notably, astronomy. As the intellectual capital of the Muslim world, which stretched from Spain to China, Baghdad attracted scholars from many different locations. Its predominant faith was Islam; however, the city became a melting pot of many other religions and cultures. For a time, Baghdad had a relatively open and tolerant society that allowed the city to flourish. The House of Wisdom was a library established in Abbasid-era Baghdad that soon grew into one of history's greatest intellectual centers. It was a hub of translation, philosophical exchange, and innovation.

Today, Baghdad serves as Iraq's capital, and some estimate it to be the Arab world's third-largest city by population after Cairo and Riyadh. Tragically, the city has suffered many deaths, infrastructure damage, and a loss of irreplaceable historical artifacts due to recent conflicts and instability. It is among the most dangerous cities on Earth, and travel there is not recommended because of the risks of terrorism and armed conflict that plague the area. Today's Baghdad is about as far as one can imagine from the city during its golden age when the urban center was a beacon of peace, tolerance, and scholarship. Literacy rates in the city may have been higher than those in many European cities at the time.

A small hamlet among several villages along the Tigris River first bore the pre-Islamic name Baghdad. The abundant water source has sustained human settlement in the area for millennia. In the eighth century, the Abbasid dynasty (the second Muslim dynasty) founded its capital in the propitious riverside location where the preexisting settlement of Baghdad stood. The meaning of "Baghdad" is disputed; however, many scholars think it means "God-given," and is of Persian origin. During the Abbasid era, Baghdad's official name was the City of Peace (*Madinat al-Salam*).

The first Abbasid caliph, or ruler, Al-Mansur, summoned engineers, architects, surveyors, and artists from many countries to construct the city over four years (764–768). The city's construction began in July, as demanded by the Abbasid court astrologers. Those astrologers believed that starting construction under the Greek astrological sign of Leo, the lion, would ensure the city's success.

Although the city's origin may have been dictated by astrology, a pseudoscience, there was not always a stark distinction between astrologers and astronomers. People who believed that the night sky could foretell human destinies were highly motivated to accurately predict the stars' movements. Hence, many astrologers studied legitimate astronomy and for centuries, many people considered astrology to be a branch of astronomy.

If you could visit Baghdad during its golden age, you would have entered a hectic commerce and scholarship center teeming with people of many different cultures speaking various languages. Many Baghdadis would have worn sandals and luxurious garments combining elements of the Arab, Irano-Turkic, and Hellenistic Mediterranean styles of dress. At the center of the city's circular layout—defined by rounded archways and curving walls—rose the domes of the caliphal Palace of the Golden Gate and the city's main mosque. The ninth-century author Al-Jahiz wrote:

> I have seen the great cities, including those noted for their durable construction. I have seen such cities in the districts of Syria, in Byzantine territory, and in other provinces, but I have never seen a city of greater height, more perfect circularity, more endowed with superior merits or possessing more spacious gates or more perfect defenses than [Baghdad].

Baghdad was abuzz with commerce. Alongside the city's four main roads, positioned like spokes on a wheel within the city's circular design, stood vaulted arcades where merchants conducted their trade. In the crowded chaos of the city's famous bazaars, you would have found goods from around the world, delivered by caravans of camels traveling the Great Khurasan Road to the city or arriving via the Tigris River trade route. You would have seen fine silk and pottery from China, elephants and spices from India, as well as rubies and other precious stones from what is now Sri Lanka, and local delicacies, such as *judhaba*. (Medieval Baghdadis were passionate about food, with the city's leaders holding elite cooking competitions.) Horrifyingly, you also would have seen people for sale—Baghdad practiced slavery, as did all major societies at that time.

In the bazaars, you would also find astrologers offering their services and many objects for sale decorated with artistic depictions of the planets and the Greek zodiac constellations. But there was more to the city's connection to the night sky than a popular enthusiasm for astrology.

In the House of Wisdom or Grand Library of Baghdad, you would have seen astronomers hard at work, occupying a prominent position alongside other scholars. Adding to the city library's collection of books and manuscripts became a point of pride for the city's rulers. By the ninth century, the city housed an immense amalgamation of writings composed in Persian, Syriac, Sanskrit, Greek, and other languages, and scholars produced Arabic translations of those works. Baghdad's scholars' large-scale translation effort has come to be known as the "translation movement," sometimes called the Graeco-Arabic translation movement, due to its emphasis on translating Greek wisdom.

The caliph Al-Ma'mun, who reigned from 813 to 833, allegedly paid one particularly acclaimed translator, Hunayn ibn Ishaq (808–873), the weight of every book that he translated in gold. He felt that wisdom was, quite literally, worth its weight in gold. Hunayn ibn Ishaq, nicknamed Sheik of the Translators (sheik being a title for a prince or ruler), became the era's most prolific decipherer of Greek medical and scientific texts. He was a Christian, and his ability to achieve a high social position despite being part of a religious minority testifies to the cosmopolitanism and tolerance of the era. His son, Ishaq ibn Hunayn (c. 830–c. 910), continued the family tradition by translating Euclid's *Elements* and Ptolemy's *Almagest* into Arabic. The city's leaders had long admired Euclid, and Baghdad's circular design may be a homage to Euclid's geometric teachings.

The *Almagest* was perhaps the first major work on astronomy. (A rival for that title might be Aristotle's *De Caelo*, written circa 350 BCE.) After the former tome's translation into Arabic, Baghdad's astronomers set about correcting several of Ptolemy's calculations regarding the planets' movements. They also perfected the astrolabe, an important tool not only in astronomy but also in navigation. Furthermore, they developed spherical trigonometry and algebra, two forms of mathematics essential to calculating the stars' movements with precision.

Muhammad ibn Musa Al-Khwarizmi, a Persian polymath, astrologer, and astronomer appointed to head the House of Wisdom in 820, invented the sine quadrant. That instrument takes angular measurements of altitude used in astronomy and navigation. In 828, Caliph Al-Ma'mun ordered the building of the first astronomical observatory in the Islamic world, within the House of Wisdom. The historian and scientist Al-Masudi— sometimes called the Herodotus of the Arabs—was born near the close of the 9th century and worked in the 10th century, and may have even invented a precursor to the telescope.

The city's openness to knowledge from foreign lands and scholars of diverse backgrounds allowed Baghdad to build on others' work and produce groundbreaking original scholarship. One House of Wisdom scholar, Abu Yusuf Ya'qub ibn Ishaq Al-Kindi (c. 800–873)—whose work spanned fields as varied as astronomy, chemistry, mathematics, medicine, metaphysics, and music—exemplified the open and tolerant worldview that allowed Baghdad to thrive. "We ought not to be ashamed of appreciating the truth and of acquiring it wherever it comes from," Al-Kindi wrote, "even if it comes from races distant and nations different from us. For the seeker of truth, nothing takes precedence over the truth, and there is no disparagement of the truth, nor belit-tling either of him who speaks it or of him who conveys it."

In that era, such broad-minded sentiments were a rarity in most places on Earth. However, they were common among the elite of Baghdad. Al-Kindi was appointed by Caliph Al-Ma'mun to serve as the tutor to the caliph's brother and eventual suc-cessor, Caliph Al-Mu'tasim, who ruled from 833 to 842. That caliph, in turn, appointed Al-Kindi as tutor to the former's son.

The prevailing interpretation of Islam encouraged phi-losophy and scientific inquiry. Several often-quoted hadiths, or sayings attributed to the Prophet Muhammad, instructed faithful Muslims to "seek knowledge." Those included an ex-hortation to "seek knowledge from the cradle to the grave" and "seek knowledge, even unto China." Those sayings were repre-sentative of the attitude held by many of Baghdad's scholars, some of whom even felt there was a religious imperative to seek out knowledge. Baghdad's scholars also believed strongly in hu-man reason and in the existence of sources of wisdom indepen-dent of divine revelation.

Unfortunately, there were also more conservative religious forces that viewed anything foreign, including foreign philosophy and scientific wisdom, as a threat to Muslim society. The conservative faction also regarded the idea of elevating human reason to the status of a source of wisdom, instead of relying solely on religious teachings for knowledge, to be blasphemous. Eventually, the triumph of the anti-rationalist and xenophobic interpretation of Islam and the subsequent persecution of the liberal Muslim scholars helped bring the Islamic Golden Age to an end.

Baghdad's ultimate downfall came in the form of conquest. It is said that the Tigris River "ran black with ink" after the Mongol invasion in 1258, led by Hulagu Khan, grandson of Genghis Khan. The Mongols demolished the House of Wisdom and allegedly threw its books in the river. Sadly, thousands of books that Baghdad collected and produced have been lost or destroyed.

But for a time, while Europe's scientific scene stagnated amid that continent's so-called Dark Ages, Baghdad's scholars made significant strides to further human understanding of the cosmos. Advances in astronomy during the later European Renaissance built largely on translations of Arabic works. To this day, the field of astronomy owes a great debt to the scholars of Abbasid-era Baghdad. Many stars maintain the Arabic names assigned to them during the Islamic Golden Age, like Altair and Betelgeuse. And today's astronomers still use the Arabic words for common astronomical terms, such as "zenith," "azimuth," and "nadir."

Baghdad during the ninth century is perhaps best known as the setting of many of the *One Thousand and One Nights* tales, widely known as the *Arabian Nights*, which were initially compiled during the Islamic Golden Age. That compendium of stories includes many well-known fables, like those of "Ali Baba and the Forty Thieves" and "Sinbad the Sailor." The tales have created an image of Baghdad in the popular imagination as a place of wonder and adventure. But in reality, the city was also the site of serious scholarly work.

For greatly advancing the field of astronomy and contributing to scholarship in several other areas, such as mathematics, early Abbasid dynasty–era Baghdad merits its place as a center of progress. Through openness to international intellectual exchange, as well as original research, the House of Wisdom and Baghdad's wider academic community made leaps that were key to many later developments in the study of astronomy. At a time when Europe was immersed in a stupor known as the Dark Ages, Baghdad had its eyes on the stars.

15

Kyoto

THE NOVEL

The next center of progress is Kyoto during the Heian (an old name for Kyoto meaning "peace") period (794–1185), a golden age of Japanese history that saw the rise of a distinctive high culture devoted to aesthetic refinement and the emergence of many enduring artistic styles. As the home of the imperial court, Kyoto was the political battleground where noble families vied for prestige by patronizing the best artists. This courtly competition produced groundbreaking innovations in many areas, including literature, and birthed a new literary form that would redefine fiction writing: the novel.

Today, Kyoto remains the cultural heart of Japan. Its well-preserved Buddhist temples, Shinto shrines, and royal palaces attract tourists from around the world, and its zen gardens have had a profound influence on the art of landscaping. Kyoto has an impressive 17 UNESCO World Heritage sites. Traditional crafts represent an important part of the city's economy, with kimono weavers, *sake* brewers, and many other renowned local artisans continuing to produce goods using heritage techniques.

In other ways, Kyoto is on the cutting edge. The city is a hub of the information technology and electronics industries, houses the headquarters of the video game company Nintendo, and contains many institutions of higher education, including the prestigious Kyoto University. The population of Kyoto now exceeds 1.45 million people, and the broader metropolitan region, including Osaka and Kobe, is the second most populated area in Japan.

Surrounded on three sides by mountains, Kyoto has been renowned for its natural beauty since ancient times, from the famous Sagano Bamboo Grove to the blossoming cherry trees along the banks of the Kamo River in the southwestern part of the city.

Archaeological evidence suggests humans have lived in the area since the Paleolithic period. Although few relics remain from the city's beginnings, some of Kyoto's architecture, such as the Shinto Shimogamo Shrine, dates to the sixth century. Japanese architecture relies heavily on wood, which deteriorates quickly, so the original building materials have not survived. However, the millennia-long Japanese tradition of continuously revitalizing wooden structures with rigorous respect for their initial form "has ensured that what is visible today conforms in almost every detail with the original structures," according to the UNESCO website "Historic Monuments of Ancient Kyoto." The most famous example of this architectural renewal is the Shinto shrine in Ise, 80 miles to Kyoto's southeast, which has been completely dismantled and rebuilt every two decades for millennia. During the Heian era, that shrine became known for imperial patronage, with the emperor often sending messengers from Kyoto to pay respects to the sacred site.

Kyoto was officially established in the year 794. Emperor Kanmu (735–806), likely feeling threatened by the growing power of Buddhist religious leaders, moved his court away from the great monasteries in the old capital of Nara. Initially, in 784, he moved the capital to Nagaoka-kyō, but a series of disasters struck after the move, including the assassination of a key imperial adviser, the death of the emperor's mother and three of his wives (including the empress), drought alternating with flooding, earthquakes, famine, a smallpox epidemic, and a severe illness that sickened the crown prince. The government's official Divination Bureau blamed that last misfortune on the vengeful ghost of the emperor's half brother Sawara, who had starved himself to death after a politically motivated imprisonment.

Although a popular narrative holds that Kanmu abandoned Nagaoka-kyō to flee the purported ghost, there may be a less spooky explanation. In 793, the emperor's adviser Wake no Kiyomaro (733–799), perhaps one of the best hydraulic engineers of the eighth century, may have convinced the emperor that flood-proofing Nagaoka-kyō would be more expensive than starting from scratch in a less flood-prone location.

Whatever the reason, in 794, Kanmu moved the capital again, erecting a new city along a grid pattern modeled after the illustrious Chinese Tang-dynasty (618–907) capital of Chang'an (see chapter 13). The lavish new capital cost a staggering three-fifths of Japan's national budget at the time. Its layout strictly conformed to Chinese feng shui or geomancy, a pseudoscience that seeks to align manmade structures with the cardinal directions of north, south, east, and west in a precise way thought to bring good fortune. The imperial palace compound, enclosed by a large rectangular outer wall (the *daidairi*), was built in the city's north and faced south. Fires presented a constant problem to the predominantly wooden complex and, although rebuilt many times, the Heian palace no longer exists. (The present Kyoto Imperial Palace, modeled on the Heian-period style, occupies a nearby location.)

From the Heian palace's main entrance emanated a large central thoroughfare, the monumental Suzaku Avenue. Over 260 feet wide, Suzaku Avenue ran through the center of the city to the enormous Rashōmon gate in the city's south. That gate lent its name to the title of the famous 1950 murder trial film by Akira Kurosawa set at the end of the Heian era. In the north of the city, close to the imperial compound, substantial Chinese-style homes housed the nobility. The emperor named his pricey metropolis *Heian-kyō* (Capital of Peace and Tranquility), now known simply as Kyoto, meaning "Capital City." (It retains that name although Tokyo succeeded it as Japan's capital in 1868.)

The Heian period of Japanese history derives its name from the era's capital city. However, the age earned its moniker's meaning and was relatively conflict free until a civil war (the Genpei War that lasted from 1180 to 1185) brought the period to a close. This long peace allowed the court to develop a culture devoted to aesthetic refinement.

For centuries, the aristocratic Fujiwara family not only dominated the politics of the court at Kyoto (marrying into the imperial line and producing many emperors), but also sought to steer the city's culture, prioritizing art and courtly sophistication. The nobility competed to fund all manner of artworks, gaining prestige from association with the era's greatest innovators in areas such as calligraphy, theater, song, sculpture, landscaping, puppetry (*Bunraku*), dance, and painting.

The nobility also produced art themselves. "The best poets were courtiers of middling rank," noted Princeton University Japanese literature professor Earl Roy Miner. "The Ariwara family (or 'clan'), the Ono family, and the Ki family produced many of the best poets" despite the Fujiwara family's greater wealth and influence. The poet Ono no Michikaze (894–966), for example, is credited with founding Japanese-style calligraphy.

It was in Kyoto that the court gradually stopped emulating Chinese society and developed uniquely Japanese traditions. For example, the Japanese *Yamato-e* painting tradition—noted for its use of aerial perspective and clouds to obscure parts of the depicted scene—competed with the Chinese-inspired *kara-e* painting tradition.

Perhaps above all, the Heian courtiers prized poetic and literary achievement. According to Amy Vladeck Heinrich, who directed the East Asia Library at Columbia University, "A person's skill in poetry was a major criterion in determining his or her standing in society, even influencing political positions." That was for good reason, as poetry played a large role in both courtly romance and diplomacy, with formal poetry exchanges strengthening the ties between potential paramours as well as other kingdoms.

The chief poetic form was the *waka*, from which the now better-known haiku was derived. Waka consist of 31 syllables, arranged in five lines, usually containing 5, 7, 5, 7, and 7 syllables, respectively. One of the era's greatest poets was the Kyoto courtier Ki no Tsurayuki (872–945), cocompiler of the first imperially sponsored poetry anthology and author of the first critical essay on waka. "The poetry of Japan has its roots in the human heart and flourishes in the countless leaves of words," he wrote. "Because human beings possess interests of so many kinds it is in poetry that they give expression to the meditations of their hearts in terms of the sights appearing before their eyes and the sounds coming to their ears. Hearing the warbler sing among the blossoms and the frog in his fresh waters—is there any living being not given to song!" (The Japanese word for song can also mean poem.)

A favorite subject for Kyoto's artists and writers was nature, especially as it changed with the seasons. As the Metropolitan Museum of Art puts it, "Kyoto residents were deeply moved by the subtle seasonal changes that colored the hills and mountains surrounding them and regulated the patterns of daily life."

Another recurrent theme was the impermanence of beauty and transience of life. Life in Kyoto was, after all, extremely short despite its relative opulence. Japanese historian Kiyoyuki Higuchi has written:

> Actual living conditions in and around the imperial court were, by today's standards, unimaginably unsanitary and unnatural. According to books on the history of epidemic disease and medical treatment, aristocratic women, on average, died at age 27 or 28, while men died at age 32 or 33. Of course, that average was

pulled down by the extremely high rates of infant and child mortality. In addition to the infant mortality rate being extremely high, the rate of women dying at childbirth was also high. . . . Looking at the specific causes of death at the time, tuberculosis (possibly including pneumonia cases) accounted for 54 percent, beriberi for 20 percent, and diseases of the skin (including smallpox) for 10 percent.

One of the period's most iconic poems, by Ono no Komachi (c. 825–c. 900), a courtier famed for her beauty, focuses on the fleeting nature of her looks:

花の色は	*Hana no iro wa*	The flower's color
うつりにけりな	*utsuri ni keri na*	already faded away
いたづらに	*itazura ni*	so meaninglessly
わが身世にふる	*waga mi yo ni furu*	I've aged, passing through the world
ながめせしまに	*nagame seshi ma ni*	gazing blankly at the rain

The poem exemplifies wordplay, and its multiple puns make it impossible to precisely translate—as the verb *furu* can mean either "to age" or "to rain," and the word *nagame* can mean either "lengthy rain" or "vacant gaze."

When Kyoto was founded, Japanese was usually written using the Chinese writing system, which was not ideal. Chinese characters could not easily convey aspects of the Japanese language that were not present in Chinese. But in the ninth century, in Kyoto, the court women—discouraged from studying Chinese—developed a simplified phonetic syllabary writing system better suited to the nuances of the Japanese language. Their system, *hiragana*, not only helped spread female literacy, but also gave writers far more flexibility and resulted in much of the best writing of the era being done by women. Today, Japanese is written using a combination of Chinese characters (*kanji*), *hiragana*, and *katakana* (another simplified syllabary developed by monks).

Perhaps the best example of the feminine influence on Heian-period Japanese literature is the competition between two of the wives of Emperor Ichijō (980–1011)—Empress Teishi (977–1001) and Empress Shōshi (988–1074)—who each sought to outdo the other and place her own son on the throne. They fought not with violence but with the arts: each tried to fill her household with superior poets and artists, thus heightening her relative prestige at court.

These dueling empresses brought about a literary rivalry for the ages between two noblewomen in their service, who went by the pen names Sei Shōnagon (c. 966–c. 1025) and Murasaki Shikibu (c. 978–c.1014). Shōnagon was a lady in waiting to Empress Teishi, and Murasaki was a lady in waiting to Empress Shōshi. Each may have been summoned to serve her respective empress specifically because of her literary talent.

In the year 1002, Shōnagon completed *The Pillow Book*, a compilation of poetry, observations, and musings now deemed a masterpiece of classical Japanese literature and among the best sources of information on Heian court life. Murasaki fired back with a masterpiece of her own and wrote scathing critiques of Shōnagon's writing and personality. By the year 1008, at least part of Murasaki's *The Tale of Genji* was in circulation among Kyoto's aristocracy.

The Tale of Genji—which chronicles the youth, romances, and eventual death of a handsome and frequently lovestruck prince—is often considered the world's first novel. The *Encyclopedia Britannica* notes *The Tale of Genji* remains "the finest work not only of the Heian period but of all Japanese literature and merits being called the first important novel written anywhere in the world."

The Tale of Genji contains many of the elements that define novels to this day: it was a lengthy prose fiction piece with a central character and minor characters, narrative events, parallel plots, and, of course, conflict. The novel also features about 800 wakas, which the characters often use to communicate. The story became an immediate hit among the nobility, inspiring numerous paintings of the novel's scenes.

While the novel's focus is an idealized vision of courtly love, it also contains untimely deaths and other unpleasant details that would have been all too familiar to Kyoto's courtiers. For example, there is no mention of bathing in *The Tale of Genji*, which sadly reflected Kyoto's state of hygiene. As Higuchi points out:

> The custom of bathing was not widespread among the nobility of that time. . . . Although beyond the imagination of people today, if a Heian noblewoman were to approach you, her body odor would likely be powerful. Moreover, whenever they caught colds, they would chew on raw garlic, increasing the odor level even more. A passage in *Genji* clearly illustrates this point: a woman writing a reply to a man asks that he please not stop by tonight since she reeks from eating garlic.

Kyoto's greatest literary feud had a decisive victress. Shōnagon remains relatively unknown outside of Japan, and the empress she served died during childbirth in her early 20s.

Murasaki's writing has gone down in history, and the empress she served lived to see two of her sons become emperors. Today, an entire museum dedicated to *The Tale of Genji* stands in Uji just outside Kyoto.

The Heian period came to a close with the rise of the samurai (hereditary military nobility) culture, and the de facto rulership of Japan transferred from Kyoto's refined albeit unbathed courtiers to warring military generals called *shoguns*.

To this day, the Japanese imperial family still runs an annual poetry-writing contest. Whereas in the Heian era typically only the nobility and monks had the time and education to compose poetry or prose, today, amateur writing is a popular pastime throughout Japan and the rest of the developed world.

Many centuries after Kyoto's era of literary brilliance, in 1905, American professor of English Selden Lincoln Whitcomb opined, "The novel is the most comprehensive form of representative art that man has discovered." For being at the center of the novel's invention, a turning point in the history of the literary arts, and its numerous other achievements in art and poetry, Heian-era Kyoto is rightly a center of progress.

16

Bologna
UNIVERSITIES

Our next center of progress is Bologna, home to the first university (as commonly recognized) and the oldest continuously operating university in the world today. What constitutes the first university is, of course, contentious. UNESCO has claimed that Nalanda—a Buddhist monastic center of study in ancient Magadha (in what is today Bihar, India) founded during the Gupta Empire in the fifth century—was the first residential university. The *Guinness Book of World Records* recognizes the madrasa of Al-Karaouine, founded as a mosque in the year 859 in Fez, Morocco, as the earliest university. But the University of Bologna, traditionally said to be founded in 1088, was the earliest institution to award degrees and promote higher learning in the manner of a modern college or university.

Today, Bologna is the seventh most populous city in Italy and home to more than 1 million people. The city's symbol is *Le Due Torri* (the Two Towers), stone structures that may date to 1109 and 1119, respectively. (A scarcity of documentation from that period means the exact construction dates remain a bit of a mystery.) Despite sustaining damage from bombing during World War II, Bologna's historic city center has remained largely intact and, at 350 acres, is Europe's second-largest stretch of medieval architecture. The major historic squares are dominated not by statues of generals or political figures, but by tombs and memorials to medieval professors. Although less popular with tourists than Florence, Venice, or Rome, Bologna has a burgeoning tourism industry. Other prominent local industries include energy, machinery, the refinement and packaging of local agricultural products, fashion, and automotives. The city is the headquarters of both Ducati, a motorcycle company, and Lamborghini, which produces luxury sports cars.

The city has three nicknames. *La Rossa* (the Red) is for its stunning medieval architecture, defined by red rooftops and lengthy UNESCO-protected red terra-cotta porticoes that make it possible to traverse much of the city while remaining in the shade. (Bologna also has a reputation for left-leaning politics, giving that nickname a double meaning.) *La Dotta* (the Learned) is for its long tradition of devotion to knowledge and for its many university students, as well as its status as the city that produced the first university. And *La Grassa* (the Fat) is an acknowledgment of the city's culinary innovations and reputation as one of Italy's gastronomic capitals.

Bologna's contributions to global food culture are significant. The city lends its name to Bolognese sauce, a meat-based pasta sauce popular in Italian cuisine that dates to at least the 18th century. Its variations are served in Italian restaurants around the world. But the city is perhaps most famous in the English-speaking world as the origin of the processed lunch-meat known as bologna—with bologna corrupted into the pronunciation baloney rather than ba-loan-ya—or simply spelled baloney. (Either spelling is acceptable for the food.)

Bologna is a variation of Bologna's mortadella sausage, which may have originated as long ago as the 14th century. Both mortadella and bologna are made of ground heat-cured pork. Italian immigrants to the United States popularized bologna in the early 20th century. An inexpensive product made from scraps of leftover pork, in America, "baloney" has also come to mean "nonsense." That is ironic given that, far from encouraging nonsense, the city of Bologna spearheaded humanity's search for truth through higher education.

Bologna enjoys a prime location amid broad fertile lowlands next to the Reno River—to this day one of Italy's leading agricultural regions. It is thus unsurprising that Bologna was first inhabited as early as the ninth century BCE.

The city's desirable location meant it was frequently conquered by outsiders. The original Etruscan city of Felsina (as Bologna was then called), fell to the Gauls by the fourth century BCE. A Celtic people, they called the settlement *Bona*, meaning "fortress." In 196 BCE, Bona became a Roman outpost bearing the Latinized name Bononia, from which Bologna is derived. After the fall of the Roman Empire, Bologna was repeatedly sacked and variously occupied by invading Visigoths, Huns, Goths, and Lombards. The city was then conquered by the Franks, led by King Charlemagne, in the eighth century. Hungarians sacked the city in the 10th century.

By the 11th century, Bologna sought to escape feudal rule and become a free commune, with the motto *Libertas* (Freedom). Exactly when Bologna made the transition is unknown, but the oldest surviving constitution of the city dates to 1123. However, the city did not remain independent for long, as various warring nobles of the Italian medieval and Renaissance periods vied for control of the city.

Limited medieval records make dates uncertain and the precise order of events unclear. However, at some point during the 11th century, Bologna became the center of revived interest in higher education, particularly the study of law. Lay students from across Europe flocked to Bologna to study law under a renowned jurist known as Pepo, an expert on Justinian the Great's compilations of Roman law.

Upon their arrival, foreign students were faced with discriminatory city laws. Bologna allowed collective punishment— the charging of any foreigners with the crimes and debts of their compatriots. The city could, in other words, seize a Frenchman's property to pay another Frenchman's debt, and punish a Hungarian for a crime committed by a different Hungarian. Because Italy was not yet a unified political entity, many groups who are today Italian, such as Sicilians, counted as foreign nationals and were also subject to collective punishment in Bologna.

Bologna's growing body of foreign students decided to try to change the laws concerning collective punishment that made living in the city perilous for nonnatives. They formed a guild, a kind of mutual aid society, known as the *universitas scholarium*. The guild hired legal scholars to give organized instruction to the students, and the latter successfully petitioned the Holy Roman emperor Frederick I (1122 or 1123–1190) to aid their cause. Frederick I issued a charter officially recognizing the University of Bologna. Known as the *authentica habita*, the charter granted Bologna's foreign scholars protection from collective punishment and gave them the right of "freedom of movement and travel for the purposes of study." The word *universitas*, which meant "guild" in Late Latin, was coined to describe the organization and gave us the modern sense of the word "university."

Like today's universities, the University of Bologna developed separate departments for different fields of study, such as theology, law, medicine, and philosophy. And like today's universities, the University of Bologna set degree requirements and awarded bachelor's, master's, and doctoral degrees. By pioneering the university model of instruction, the University of Bologna helped humanity make progress in many areas—but especially legal studies. Pepo is often said to be the first university legal instructor.

Pepo was soon far surpassed by his student Irnerius (c. 1050–after 1125), who also went on to teach at the University of Bologna. He was originally a student of rhetoric and didactics. His wealthy patroness, one of Italy's most powerful nobles at the time, Matilda of Tuscany (c. 1046–1115), convinced him to switch fields and study jurisprudence. Nicknamed *Lucerna Juris* (Lantern of the Law), Irnerius's scholarship is credited with creating much of the medieval Roman law tradition. His glosses on the ancient Roman law code helped move medieval law, which was sometimes disordered and contradictory, in the direction of becoming more systematic and rational like the ancient Roman legal system. Irnerius's most famous students—Bulgarus, Martinus Gosia, Hugo da Porta Ravennate, and Jacobus de Voragine—came to be called the Four Doctors of Bologna. Each allegedly had a different approach to legal philosophy.

By the end of the 12th century, the University of Bologna was widely recognized as Europe's premier center of higher learning, particularly legal studies, drawing an ever-larger crowd of elite international students from across the continent. Thomas à Becket (c. 1120–1170), a famed Archbishop of Canterbury who sought to preserve the independence of the Church from the state, and who is now revered as a martyr-saint in both the Catholic and Anglican Churches, studied law at the University of Bologna in his youth. The Florentines Dante Alighieri (c. 1265–1321) and Francesco Petrarca, commonly Anglicized as Petrarch (1304–1374), also both studied at the University of Bologna. Other famous alumni include four former popes. Yet another renowned alumnus was the Dutch humanist Erasmus of Rotterdam (c. 1469–1536), an early champion of religious toleration and peace, and arguably a hero of progress.

From the 12th to the 15th century, the university had between 3,000 and 5,000 students. Today, the university has over 90,000 students.

The University of Bologna is also commonly said to be the first university to award a degree to a woman and allow one to teach at the university level. According to tradition, in 1237, a noblewoman named Bettisia Gozzadini (1209–1261) graduated after studying philosophy and law and began lecturing on jurisprudence in 1239.

Whether Gozzadini actually graduated from Bologna became a point of contention in the 1700s. The law student Alessandro Machiavelli (1693–1766) sought to provide evidence (possibly faked) of Gozzadini's achievement to support the Bolognese Countess Maria Vittoria Delfini Dosi's request to be granted a law degree. Despite Machiavelli's efforts, the countess' request was ultimately denied. Male scholars who opposed the idea of granting women degrees sought to dismiss Gozzadini as a popular legend. Scant records from the medieval period make the truth hard to discern.

That said, the University of Bologna employed the first salaried female university professor, the physicist Laura Bassi (1711–1778). She is credited with popularizing Newtonian mechanics in Italy. She was also the first woman to earn a doctoral degree in science and only the second woman to receive any doctoral degree. Bassi's doctorate was also from the University of Bologna.

Bologna boasts many achievements in realms as diverse as architecture and gastronomy. But creating the world's first university has been Bologna's defining contribution to human progress. Universities have helped promote scholarship, innovation, and higher learning ever since. By promoting the study of the law, in particular, Bologna helped humanity in its pursuit of an improved system of justice.

The translated university motto reads, "Saint Peter is everywhere the father of the law; Bologna is its mother." The university's full name is L'Alma Mater Studiorum— Università di Bologna (the Nourishing Mother of Studies— University of Bologna). From that name, we get the term "alma mater," popularly used by university graduates throughout the world to refer to whatever university they attended. But the mother of all universities is Bologna. For birthing the modern university system, medieval Bologna is rightly a center of progress.

17

Hangzhou

PAPER CURRENCY

Our next center of progress is Hangzhou in 12th-century China, during the late Song dynasty's so-called premodern commercial revolution or period of proto-industrialization. With its innovations in printing and manufacturing, the book *The Earth and Its Peoples: A Global History* even claims the "Song came closer to initiating an industrial revolution than any other premodern state." The Song dynasty, which spanned from 969 to 1276, was a time of dynamism and invention. Through trade and industry, the Song empire became the richest on Earth. The dynastic capital, Hangzhou, was the wealthiest and most populous city in the world. Song-era China became the first country to print paper money, which is far easier to carry in large amounts than metal coins. Hangzhou served as a money-printing center and a hub of innovation and creativity.

During the Song era, the average Chinese person experienced extraordinary growth in income level as the economy expanded. The economy grew because of new technological and agricultural advances and efficient trade routes that produced a genuinely nationwide market. The era also witnessed a significant increase in international exchange, as Chinese merchants expanded their trade networks as far as East Africa. Growing wealth helped motivate the adoption of paper money, as people found themselves dealing with larger transactions than in the past.

Today, Hangzhou is one of China's top commercial bases. It is also the southern terminus of the Grand Canal, which is the world's longest artificial river and a UNESCO World Heritage site. True to its rich history of innovation, Hangzhou continues to serve as a hub of enterprise. Hangzhou houses the headquarters of various internet enterprises such as e-commerce giant Alibaba and is a growing technology center. Hangzhou is the heart of the "Hangzhou metropolitan area," China's fourth-largest metropolitan area by population, and is home to some 20 million people. Hangzhou is also a popular tourist destination within China. Hangzhou maintains many well-preserved cultural sites showcasing the city's history. It even has a large history-based theme park—Song Dynasty Town, or Songchen—filled with costumed reenactors portraying residents from the city's golden age.

The Italian explorer Marco Polo famously described Hangzhou as "the most beautiful and magnificent city in the world" and called it the "City of Heaven" during a visit in the 13th century. That was after the Song dynasty had ended; however, much of the architecture and wealth of the city that Marco Polo observed was nonetheless a legacy of that era. (A statue of Marco Polo stands prominently in a lakeside park in the city, admiring Hangzhou's beauty to this day.) A common Chinese saying echoes Marco Polo's sentiment: "Above, there is Heaven; below, there are Hangzhou and Suzhou"—the latter being another beautiful city just to Hangzhou's north.

Hangzhou has been an important city since the seventh century when its Grand Canal was first built to connect the urban center to Beijing. Today, the canal remains the main north-south waterway in China. But the city's golden age began when the Song dynasty made it its capital. The Song era saw the rapid adoption of woodblock printing, a technology that supercharged intellectual life. Hangzhou ranked first in China when it came to both volume and quality of woodblock printing. The technique—which consisted of carving text and pictures into wooden blocks, covering them with ink, and pressing the blocks against paper—provided a way to mass-produce books, documents, and banknotes.

Woodblock printing developed in Buddhist monasteries to reproduce spiritual texts, with examples dating as far back as 200 CE, and the method was well established by the ninth century. However, it was during the Song era that woodblock printing was first widely adopted for nonreligious purposes. In the 11th century, the artisan and inventor Bi Sheng (990–1051) devised movable type. The adoption of printing technology dramatically lowered the cost of books and encouraged the spread of literacy. Widespread printing not only led to a veritable tidal wave of artistic output such as poetry and dramatic texts, but also sped up scientific progress—for example, by aiding the dissemination and advancement of pharmacological and medical knowledge.

If you could visit Hangzhou during its golden age, you would enter a gorgeous metropolis bursting with art, commerce, innovation, and a spirit of openness. The crowds would have been formidable; by the end of the Song era in 1276, Hangzhou was home to about 1.75 million residents according to some estimates. That is slightly more than the current population of Phoenix, Arizona, but it represented an unprecedented urban concentration of people. Although poor by modern standards, the city's people were then the richest on Earth. Looking out onto the harbor, you would have seen large multisectioned ships with up to four decks and a dozen sails at a time when Europeans still traveled in tiny galleys powered chiefly by the muscle of rowers.

Thanks to advances in dyeing and weaving and textile industry developments, the city's people would have worn a wide variety of beautiful and luxurious robes. You would not have seen many high-ranking women walking around. Despite the era's many advances, it was also the beginning of "foot-binding" among China's elite. That cruel practice consisted of repeatedly breaking the bones in women's feet, starting in early childhood, to contort feet into an unnatural shape that was considered beautiful but made walking physically painful.

In the marketplace, you would see a food culture emerging that has since come to define Chinese cuisine. During the earlier Tang dynasty (the golden age of another center of progress featured in these pages, Chang'an), China's dominant grains were wheat and millet, and the most common drink was wine. During the Song dynasty, rice and tea became the country's staple food and beverage and have remained so to this day.

You would have been astonished by the city's elaborate architecture. (China's traditional upturned roofs originated in the

Song dynasty.) Hangzhou's striking temples, many of which still stand today, were a testament to the era's philosophical and spiritual diversity. As writer Eric Weiner put it, "The blending of Buddhist and Confucian thought yielded a remarkably tolerant atmosphere." Different thought systems coexisted and thrived. Conversation rose to an art form, and as the city became wealthier, art of all kinds became an important part of everyday life. In previous eras, poetry was limited to religious subjects; however, in the Song era, poetry expanded to deal with every topic imaginable, and poetry competitions were frequent.

Hangzhou was the site of great creativity. In the 11th century, the polymath Shen Kuo (1031–1095) invented the magnetic compass. He also drew the world's first topographical map and was the first person to record the process of sedimentation. Shen's surviving notebooks have garnered comparisons to Leonardo da Vinci's for their breadth. Shen's work spanned topics such as mathematics, astronomy, meteorology, geology, zoology, botany, pharmacology, agronomy, archaeology, ethnography, cartography, diplomacy, hydraulic engineering, and finance. Shen was also a prolific poet.

Another intellectual of the Song era was Su Tung-Po (1037–1101). He was once a governor of Hangzhou but is better known for his art, work as an engineer, and insightful poetry. Su Tung-Po's poetry reveals a self-effacingly unflattering view of government officials:

> Families when a child is born
> Hope it will turn out intelligent.
> I . . . only hope that the baby will prove
> Ignorant and stupid.
> Then he'll be happy all his days
> And grow into a cabinet minister.

Given that many of the city's advances came from the private sector, Su Tung-Po's attitude was understandable. Even paper money was arguably a private-sector invention. As early as the Tang dynasty (618–907), the impracticality of transporting strings of heavy coins inspired Silk Road merchants to use paper promissory notes instead to make purchases. (Chinese coins had square holes in the middle to allow for stringing them together.) Private agents originally produced those notes. At the beginning of the Song dynasty, the government recognized the value of that innovation and licensed deposit shops where people could exchange coins for such promissory notes, thus somewhat standardizing the system. Then, in the 12th century, the government gave still greater recognition to the concept of paper money by issuing the first official paper currency, called *jiaozi*. Those banknotes often featured intricate illustrations of commerce.

During the golden age of Hangzhou's "economic revolution," the Song leaders managed to largely avoid international conflict by defusing tensions with trade agreements and tributary offers. Thus, Hangzhou was mostly at peace during its peak years, leaving its residents free to engage in enterprises that further enriched the city. According to American historian Philip D. Curtin: "Between . . . 960 and . . . 1127, China passed through a phase of economic growth that was unprecedented in earlier Chinese history, perhaps in world history up to this time. It depended on a combination of commercialization, urbanization, and industrialization that has led some authorities to compare this period in Chinese history with the development of early modern Europe six centuries later."

Factories situated in Hangzhou and the other major Song-era cities of Chengdu, Huizhou, and Anqi printed paper money with a uniform design using woodblocks and six different ink colors. Each city used multiple banknote seal stamps and different fiber mixes in the paper currency they produced to make counterfeiting difficult. In 1175, as many as 1,000 employees may have worked in Hangzhou's paper money factory each day. The earliest money notes expired after just three years (and could be exchanged for their nominal value in copper coins), and their use was limited to certain regions of the Song empire. Then, in 1265, Hangzhou's factories printed the first truly national currency. That currency exhibited a unified design and was accepted across the empire, and its value was backed by silver or gold. The paper money notes were available in various denominations. Unfortunately, that national currency was only used for nine years before a Mongol invasion ended the Song dynasty.

The concept of paper currency proved more lasting than the Song dynasty that created it. The subsequent Mongol Yuan dynasty issued its own paper currency, known as the *chao*. However, the Mongols did not tie their currency's value to anything and printed more and more banknotes until runaway inflation degraded the currency's worth. Paper money can be susceptible to hyperinflation without sound monetary policy. Paper currency has nonetheless proved to be a lasting and practical invention that is now used worldwide.

For being a hothouse of invention and creativity and the site of an early economic revolution that gave the world paper money, 12th-century Hangzhou is deservedly a center of progress. Bolstered by printing technology and paper currency's relative efficiency, the Song era saw a steady stream of technological breakthroughs. Those included the compass, the first mechanical clocks, and the invention of forensic science. The economic and technological advancements of the Song era translated into improving living conditions for the average person. By practically every measure of human well-being, ranging from sanitation to literacy to average income, China was superior to Europe in the 12th century. Thanks to relative peace, far-ranging trade, and cultural openness, Hangzhou prospered and produced many accomplishments, including inventions that we still use today.

18

Florence

ART

Perhaps no city so perfectly exemplifies the idea of progress as Florence during the Renaissance. Known as "the Jewel of the Italian Renaissance" and sometimes, "the birthplace of the Renaissance," Florence was at the heart of too many groundbreaking developments to mention. The city contributed to significant advances in politics, business, finance, engineering, science, philosophy, architecture, and—above all—artistic achievement. Florence produced historic art projects throughout the Italian Renaissance (1330–1550), particularly during the 15th century, the city's golden age.

The Florentines' wide-ranging contributions to human progress are all the more amazing when one considers that a pandemic killed half the city's population of around 85,000 people in the 14th century, including the painter Bernardo Daddi (1290–1348). (For more detail on the pandemic, see the following chapter.)

Today, Florence is the capital city of the Italian region of Tuscany. Tuscany, known for its natural and architectural beauty, may be the most frequently photographed region in Italy. Florence is also Tuscany's most populous city, with more than 300,000 inhabitants and 1.5 million residents in its greater metropolitan area. With its long history and arresting scenery, Florence is a popular tourist destination that often merits a place on lists of the world's most beautiful cities. The Historic Center of Florence is a UNESCO World Heritage site. Florence is also a key center of Italy's fashion industry.

That is fitting because the story of Florence's rise to prominence began with cloth. More precisely, woolen cloth. Tuscany has plenty of sheep and grazing land, and for centuries, Florence produced wool locally. But circa 1280, Florentines began to import wool from England, as English wool was of higher quality. Florence's river, the Arno, made cleaning large amounts of imported wool achievable.

Florence enjoyed a central trade location between East and West. Some Florentine merchants realized that their city was perfectly situated to combine top-notch wool from England with the world's best dyes from Asia—resulting in uniquely luxurious woolen cloth. The Florentine woolen fabric was soon in high demand throughout Europe. By the 14th century, one-third of Florence's population worked in the woolen cloth industry.

Thus, international trade set Florence on a path toward success in the fabric business. The city's booming cloth industry created a large, wealthy merchant class. As the Florentines grew rich, new finance and banking innovations further elevated the city's prosperity.

As Florence's wealth increased, its people needed to exchange larger and larger amounts of florins (i.e., the city's currency from 1252 to 1533). Thus, Florence became the first city in centuries to mass-produce gold-coin currency. Florentine bankers soon became renowned experts at coin valuation, and the florin became the most trusted currency in Europe.

Moreover, Florence became the first city-state whose bankers charged interest on loans. Historically, most bankers throughout Europe would not charge interest because doing so was widely considered to be a sin called usury. However, giving out loans without charging interest is risky and usually unprofitable. As a result, for many years, Jews were among the only Europeans who could enter the money-lending business without going bankrupt. Jewish law did not prohibit lending money to others at interest, provided they were gentiles (non-Jews). Unfortunately, an anti-Semitic trope arose related to this practice, appearing in William Shakespeare's play *The Merchant of Venice*, for example. But Florence's Christian bankers found a loophole with a bit of creative accountancy: they presented interest as a voluntary gift on the part of borrowers or as compensation for the risk taken on by lenders. (Those who failed to pay the technically voluntary fees were often blacklisted by Florence's banks and unable to obtain future loans.)

Charging interest let the Florentine bankers make credit widely available in a profitable and thus sustainable way. Not only did that put loans within reach of many Florentines, but

Florence's bankers soon became the moneylenders of choice for the wealthy and powerful throughout Europe, including royalty and the pope. The bankers' financial services also included facilitating trade by furnishing merchants with bills of exchange that allowed them to pay off their debts while in a different town from their creditors—a concept familiar to anyone who has ever mailed a modern check. Florence's banks accomplished that by opening offices or branches in various cities. Florentine bankers also perfected double-entry bookkeeping.

Through its lucrative cloth trade and innovative banking industry, Florence quickly rose to become the wealthiest city in Europe during the Renaissance. That wealth improved the lives of everyday people throughout the city. For example, Florence became the first city in Europe to pave its streets, in 1339.

The city's wealth resulted in more than improving material conditions—it also prompted a shift in the way people thought. Humanism and classicism came into vogue. Humanism was an intellectual movement focused on human achievement and the enjoyment of life's pleasures, such as beautiful gardens and art. Humanism contrasted starkly with the prior widely held belief in asceticism. Florence's growing upper and middle classes increasingly engaged in intellectual pursuits, such as studying history and classical Roman texts, which allowed the former to recover lost knowledge in many fields. It is fitting that the literal meaning of Renaissance is "rebirth." For example, by studying old Roman writings, the artist Raphael (1483–1520) managed to recreate a rare blue paint pigment invented by ancient Egyptians.

Florentines considered their city to be the "New Rome." That was partly because they brought back into practice much of the knowledge of the ancient Romans that had fallen into disuse. Like the ancient Romans, the Renaissance Florentines also felt that their home embodied an ideal city-state republic, guaranteeing individual freedom and the right to political participation to a portion of the population. Although, like Republican Rome, Florence was not a true democracy but an oligarchy. The republic had slavery, like most societies at the time, and was also infamous for political intrigue.

Florence's relatively inclusive political system, classicist appetite for knowledge, humanist zest for life, and the freedom afforded by growing prosperity all combined to give rise to the ideal of the "Renaissance man." Many Florentine men strived to attain far-ranging expertise across fields such as art, literature, history, philosophy, theology, natural science, and law. The educator Pietro Paolo Vergerio (c. 1369–1444), who studied in Florence among other cities, wrote the era's most influential educational treatise. That treatise, *On the Manners of a Gentleman and Liberal Studies*, published in 1402 or 1403, helped create the concept of a well-rounded liberal arts education.

Florence was the first Italian city-state to host a center for learning—the University of Florence. It was established in 1321 and relocated to nearby Pisa in 1473. The scholar Giovanni Boccaccio—today best remembered for authoring the *Decameron* (a collection of stories also collectively known as *l'Umana commedia* or "the Human Comedy")—helped make the university into an early capital of Renaissance humanism. Together with the scholar Petrarch (1304–1374), whose rediscovery of Cicero's letters is sometimes credited as starting the Italian Renaissance, Boccaccio popularized writing in the vernacular rather than in Latin. Florence's greatest poet, Dante Alighieri (c. 1265–1321), authored his narrative poem, the *Divine Comedy*—which is still widely called the greatest Italian literary work—in the vernacular. That work was so popular throughout Italy that it helped establish the local Tuscan dialect as the default, standardized version of Italian, replacing other regional dialects.

Although they generally received less education than men overall, Renaissance women from wealthy families were educated in the classics and sometimes the arts. A notable example was Sofonisba Anguissola (c. 1535–1625), an Italian noblewoman who studied painting under the acclaimed artist Michelangelo (1475–1564). Although he spent most of his life in Rome, Michelangelo considered himself a Florentine (he worked in Florence in his youth). Anguissola attained professional success, and became the official court painter to the king of Spain. Her achievements paved the way for other European women to pursue serious artistic careers.

Florence's rise was not without difficulties. In the 1300s, the bubonic plague pandemic swept through Italy. By 1348, the pandemic had reached Italy's interior, including Florence, and revisited the city in periodic bouts. As noted earlier, the illness is estimated to have killed approximately half of Florence's population. Such a widespread loss of life created intense economic and social disruption. Yet even in the aftermath of that tragedy, Florence continued to innovate and create. By the 15th century, the city had entered its golden age. The citizens poured their fortunes into patronage of the arts, and the Catholic Church also paid for many artistic projects. Pope Julius II (1443–1513), in particular, was known for artistic patronage. Florence's wealthiest banking family, the Medicis, also became famous for financially supporting Renaissance artists.

Florence was teeming with geniuses. If you could take a stroll through the city in the 15th century, you might run into the polymath Leonardo da Vinci (1452–1519). Born and raised in Florence, Leonardo was the quintessential Renaissance man, whose notebooks spanned topics ranging from anatomy to cartography and painting to paleontology. Or you might meet the artist mentioned earlier, Raphael, considered one of the three great masters of the Renaissance, together with Leonardo and Michelangelo. You might greet the sculptor Donatello (1386–1466). You might encounter a young Niccolò Machiavelli (1469–1527), who worked as an official in the Florentine Republic, wrote the famous treatise *The Prince*, and is often labeled the father of modern political philosophy and political science. You might chance upon the explorer and merchant Amerigo Vespucci (1454–1512), from whom the Americas get their name. You might pass by the art workshop run by the artist and businessman Andrea del Verrocchio (1435–1488), who mentored many of the city's best artists, including Leonardo. Verrocchio's workshop also helped cultivate Florence's atmosphere of competition in the development of new artistic techniques. You might cross paths with Filippo Brunelleschi (1377–1446), often called the first modern engineer and the father of Renaissance architecture, who designed Florence's iconic cathedral. Or perhaps you would happen upon Sandro Botticelli (c. 1445–1510), yet another Florentine artistic legend.

Whereas European art had degraded during the Dark and Middle Ages to relatively simplistic figures, the Renaissance not only resurrected the hyperrealistic and proportional sculpture style of the ancient Greeks and Romans but went further in developing extraordinarily sophisticated painting techniques.

Florentine artists perfected proportionality and foreshortening (shortening lines to create the illusion of depth). Moreover, they developed the so-called canonical four techniques of the Italian Renaissance: *cangiante, chiaroscuro, sfumato,* and *unione. Cangiante* creates the illusion of shadows using a limited color palette. *Chiaroscuro* is a method of contrasting light and dark paint to convey a sense of depth. *Sfumato* is a way of subtly blurring outlines to give the illusion of three-dimensionality. And *unione* is a color transition technique that produces dramatic effects.

In sum, the Florentine artists' processes and techniques established the basis of traditional Western painting, with their methods still in use hundreds of years later.

Florentines produced many of history's most acclaimed paintings and other artworks. Those include *The Trials of Moses* fresco in the Sistine Chapel, the *Birth of Venus, Primavera,* and *Venus and Mars* by Botticelli; the sculpture *David* and the artworks of St. Peter's Basilica and the Sistine Chapel, such as the *Creation of Adam,* by Michelangelo; *The School of Athens* (referenced in that city's chapter) by Raphael; and *The Last Supper* and *The Virgin of the Rocks* by Leonardo.

The *Mona Lisa*—an early 16th-century portrait by Leonardo depicting a Florentine merchant's wife—is today the world's most visited painting. It is housed in the Louvre in Paris, which attracts around 10 million annual visitors.

Not everyone was happy with Florence's prosperity and artistic creations. Progress is seldom without controversy. An anti-humanist, pro-ascetic backlash led by the radical friar Girolamo Savonarola (1452–1498)—who merited a mention in

Chapter 2 of this book—briefly threw Florence into turmoil. Savonarola encouraged his followers to destroy paintings, musical instruments, fine clothes, jewelry, humanist books (such as the works of Boccaccio), and other allegedly sinful possessions. Mass burnings of such objects were called "bonfires of the vanities." Savonarola's movement, sometimes considered a precursor to the Protestant Reformation, eventually got him excommunicated by the pope and executed by political opponents. The burning of Florence's so-called vanities ceased, and many of the city's artistic masterpieces survive to this day.

Innovations in trade, business, and banking helped make Florence wealthy, and the Florentines spent enormous sums toward the patronage of artists. As Eric Weiner noted, "Genius is expensive." The city's merchants and bankers were as key to Florence's flourishing as the artists they funded. In turn, those artists conducted extraordinary experiments in creativity and produced some of the world's most remarkable artistic accomplishments. The powerhouse of the Renaissance, Florence not only revived lost knowledge from Greco-Roman texts but revolutionized art in a way that would come to define Western painting. Florence is also a symbol of resilience in the face of a pandemic. For these reasons, Renaissance Florence undoubtedly deserves to be a center of progress.

19

Dubrovnik

PUBLIC HEALTH

The next center of progress is now called Dubrovnik, but was known historically as Ragusa. The picturesque port city is nicknamed the Pearl of the Adriatic for its beauty. But the city has also been called the Hong Kong of the Mediterranean for its historic embrace of personal and economic freedom and its maritime trade-based prosperity. Not only was the small city-state of the Republic of Ragusa at the forefront of freedom for its time, being one of the earliest countries to ban slavery, but the glittering merchant city on the sea was also the site of an early milestone in the history of public health: quarantine waiting periods, which were first implemented in 1377. (The practice of merely isolating the sick in a nonsystematic manner is, of course, older.) In 1390, Dubrovnik also created the world's first permanent public health office. Perhaps more than any other city, Dubrovnik can claim to have helped create the idea of public health.

Today, Dubrovnik is best known for its exquisite sights, including many historic buildings and museums. It is located in the southern Croatian region of Dalmatia, best known for the dalmatian dog breed, which existed as far back as 1375. Tourism dominates the economy. Much of the city's layout remains largely unchanged from the year 1292, with narrow winding stone-paved streets; innumerable medieval monuments, towers, and monasteries; and charming garden-surrounded villas and orange groves. The Old City region is a designated UNESCO World Heritage site, boasting well-preserved Gothic, Renaissance, and baroque architecture in the form of numerous churches and palaces. The city is often considered a major artistic center of Croatia and the site of many cultural activities, theatrical and musical performances, festivals, and museums. The city's Banje Beach is also popular, and the port of Gruž is now busy with cruise ships.

The Irish playwright George Bernard Shaw (1856–1950) claimed, "Those who seek paradise on Earth should come to Dubrovnik." Fans of *Game of Thrones* may recognize Dubrovnik as the film site that brought to life the fictional seaside city of King's Landing. But whereas King's Landing was the capital of a despotic absolute monarchy, in reality, Dubrovnik was devoted to freedom to an unusual degree from its inception and is proud to have had no king. "The city-republic was liberal in character, affording asylum to refugees of all nations—one of them, according to legend, was King Richard I (the Lionheart) of

England, who landed on the offshore island of Lokrum in 1192 on his return from the Crusades," according to the *Encyclopedia Britannica*.

Dubrovnik was a tributary city-state under Venetian suzerainty from 1205 to 1358, retaining substantial independence and growing prosperous as a mercantile power. It was during that period, in 1348, when the bubonic plague first reached the city. Within four years, the disease extinguished the lives of perhaps two-thirds of the city's citizens. And that was just the first wave. During the Black Death pandemic, periodic lulls were often followed by new outbreaks.

In 1358, Hungary pressured Venice to surrender control of Dubrovnik, and the Republic of Ragusa (1358–1808) was born. It was during the republican era that the city created the novel public health measure of quarantine, and practiced it from 1377 to 1533. Although not perfect—outbreaks of plague occurred in 1391 and 1397—the measure was nonetheless revolutionary. Other cities soon implemented similar protocols, such as Geneva in 1467.

"It should not be a surprise to find Dubrovnik at the heart of quarantine's origin-story, because the city was a seafaring supernova for much of the medieval era," notes British journalist Chris Leadbeater. An aristocratic republic with fewer than 10,000 people living within its walls and a constitution resembling Venice's, Dubrovnik was ruled by a council of merchant princes selected from the patrician families that composed about a third of the city's population. Unlike in Venice, the ranks of the nobility were never formally closed, meaning that newly successful merchant families could gain patrician status. Term limits restricted the top government official, the rector, from serving for more than a month, after which he could not seek the role again for two years. According to historian Susan Mosher Stuard, Dubrovnik also "never saw an elaborate increase in bureaucratic functions or felt the great weight of government intervention as Venetians did," opting for relatively limited government interference with the city's robust trade.

If you could visit Dubrovnik during its maritime golden years (1350–1575), you would enter a vibrant coastal city filled with stone architecture, diverse travelers speaking languages ranging from German to Turkish to Italian, and awash in art and commerce. You might have glimpsed noblewomen wearing fine jewelry, who were free to trade their jewels without male permission even in that age of extreme gender inequality, thus contributing to a lucrative export market.

The Croatian economic historian Vladimir Stipetić has noted, "Dubrovnik traded like Hong Kong, Singapore, Taiwan . . . but did so some five hundred years before . . . [and like those countries] became prosperous . . . because of [its] adopted economic policy." As a result of the city's relative economic freedom, and the resources saved by the city's disinterest in military expansionism, Dubrovnik's fleet of hundreds of merchant ships at times outnumbered those of Venice, despite the latter's boasting perhaps 10 times the population of Dubrovnik. Dubrovnik's economic expansion is also, of course, owed to the innovativeness of its people. In the 15th century, a Dubrovnik humanist, merchant, and noble named Benedetto Cotrugli (1416–1469) published *Della mercatura e del mercante perfetto* (*On trade and the perfect merchant*), which is thought to be the first work on bookkeeping in the world. It was also a trade manual advocating honesty in all dealings.

The republic mediated trade between the Ottoman Empire and what was popularly called Christendom. Located at the intersection of territories practicing Islam, Catholicism, and Orthodox Christianity, Dubrovnik maintained a policy of friendly trade with people of all faiths in an era when religious tensions were high, while internally endorsing Catholicism. The city's culture was unusually "secular, sophisticated, individualistic," and cosmopolitan for its time, according to economic historians Oleh Havrylyshyn and Nora Srzentić. During its republican era, Dubrovnik became a major center of Slavic literature and art, as well as philosophy, particularly in the 15th, 16th, and 17th centuries—earning the city the nickname the Slavic Athens. It produced notable writers, such as Elio Lampridio Cerva (1463–1520), Šiško Menčetić (1457–1527), Marin Držić (1508–1567), and Ivan Gundulić (1589–1638), now regarded as Croatia's national poet. His most famous work is the "Hymn to Freedom":

O liepa, o draga, o slatka slobodo,	O beautiful, o precious, o sweet liberty,
dar u kom sva blaga višnji nam	best gift of all the treasures
Bog je dô,	God has given us,
uzroče istini od naše sve slave,	the truth of all our glory,
uresu jedini od ove Dubrave,	the decoration of Dubrave [Dubrovnik],
sva srebra, sva zlata, svi ljudcki životi	all silver, all gold, all human lives
ne mogu bit plata tvôj čistoj lipoti.	cannot repay your pure beauty.

Despite its lack of military power and its small size, Dubrovnik's economic freedom and remarkable political and social stability helped the tiny republic survive for almost half a millennium before Napoleon conquered it in 1808. Although Dubrovnik was at times compelled to provide tribute to its more powerful neighbors to maintain political independence, the republic's citizens were proud of their relative liberty. In fact, the republic's Latin motto was *Non bene pro toto libertas venditur auro* (Liberty is not sold for all the gold in the world). The republic's flag was simply the word *Libertas* (Latin for "liberty") in red on a white background. From 1792 to 1795, Dubrovnik also issued silver coins called *libertinas*, featuring the word *Libertas* in the design's central position. Moreover, the republic was among the first European countries to abolish slavery, outlawing the slave trade in 1416. The city's governing council voted that "none of our nationals or foreigners, and everyone who considers himself or herself from Dubrovnik, can in any way or under any pretext buy or sell slaves . . . or be a mediator in such trade."

Recognizing the threat that recurring outbreaks of bubonic plague posed to their city, the people of Dubrovnik took action to preserve their trading prosperity and their very existence. Thanks to Dubrovnik's public health measures, the city managed to prevent many deaths and even achieve significant mercantile expansion during the plague period.

Bubonic plague is a bacterial disease that, when left untreated, is usually fatal within days of symptoms appearing. The bubonic plague has ravaged humanity many times and has even been found in human skeletons dating to 3000 BCE. Bubonic plague cases still occur even today. The first outbreak of the illness that was widespread enough to be termed a pandemic occurred in the sixth century, during the reign of the Byzantine emperor Justinian I. But the bubonic plague pandemic that devastated Asia, Africa, and Europe in the 14th century—named the Black Death or the Great Pestilence—proved to be the most fatal pandemic in recorded history, killing perhaps as many as 200 million people, including up to 60 percent of Europe's population.

That outbreak first emerged in western China. In just three years, between 1331 and 1334, bubonic plague killed more than 90 percent of the people in Hebei Province, which covers an area of land slightly bigger than Ireland. Over 5 million Hebeian corpses presented a preview of the deaths to come.

The scale of the devastation is difficult to imagine. The Black Death laid waste to Europe from 1346 to 1353. In 1348, the bacteria wiped out 60 percent of Florence's population. That same year, the plague reached France, and within four years, at least a third of Parisians were in the grave. The following year, the plague arrived in London and halved that city's populace. In practically every city and town, the tragedy repeated itself.

One firsthand account of the devastation notes:

> This mortality devoured such a multitude of both sexes that no one could be found to carry the bodies of the dead to burial, but men and women carried the bodies of their own little ones to church on their shoulders and threw them into mass graves, from which arose such a stink that it was barely possible for anyone to go past a churchyard.

Survivors were haunted by grief and loneliness. In 1349, the Italian writer Petrarch, who lost many companions to the plague, including his muse Laura, wrote:

> Where are our dear friends now? Where are the beloved faces? Where are the affectionate words, the relaxed and enjoyable conversations? . . . What abyss swallowed them? There was a crowd of us, now we are almost alone. We should make new friends—but how, when the human race is almost wiped out; and why, when it looks to me as if the end of the world is at hand? Why pretend? We are alone indeed.

Despite life's hardships, survival was nonetheless preferable to death, and people made a great number of innovative attempts to prevent and treat the disease that was decimating humanity. Many of those measures were tragically ineffective, such as bloodletting and avoiding baths. (Bathing was thought to expand the pores and make one vulnerable to disease.) Some measures helped a little in the prevention of illness—such as avoiding foul smells, including rotting corpses, and encouraging better home ventilation.

Famously, medieval understanding of how disease spread left much to be desired. Many assumed that the Black Death was a divine punishment for mankind's sins, giving rise to the distressing flagellant movement, and some of the brightest minds of the day at the University of Paris, when commissioned by the king of France to explain the plague, concluded that the movements of Saturn were to blame. Others blamed witchcraft. Reprehensibly, still others violently scapegoated religious minorities: "Hygienic practices limited the spread of plague in Jewish ghettos, leading to the Jews being blamed for the plague's spread, and widespread massacres, especially in Germany and Central Europe," according to public health scholars Theodore Tulchinsky and Elena Varavikova.

However, although they may not have grasped the cause of the illness, medieval people did possess the general concept of contagion. They knew that the plague disseminated from one place to another and that transmission was occurring in some way: the suspected vectors ranged from the wind to the gaze of an infected person.

Fortunately, medieval people did not need to know that the bubonic plague spreads mainly via fleas to figure out that limiting contact with people and objects from known outbreak sites was the most prudent course of action. This idea became widespread in part through the works of various physicians publishing medical pamphlets or tractates throughout Europe that may have represented "the first large-scale effort at popular health instruction in history," according to public health scholars Charles-Edward Amory Winslow and Marie Louise Duran-Reynals. The Catalan doctor Jaume d'Agramont (d. 1350 of plague), for example, advised the public against eating food from "pestilential regions," and wrote that "association with a sufferer of a pestilential disease" could cause the illness to spread from one person to another "like a wildfire." The possibility of interpersonal transmission became widely suspected, even if few guessed at the flea's role as an intermediary.

Even before the plague, Dubrovnik made several strides toward better public health. While we now take basic hygiene measures for granted, Dubrovnik was something of a medieval outlier when it restricted the wanton disposal of garbage and feces in the city in 1272. The city banned swine from city streets in 1336, hired street cleaners in 1415, and created a complete sewage system in the early 15th century. Dubrovnik's relative prosperity allowed it to offer competitive wages to draw physicians from other cities, such as Salerno, Venice, Padua, and the home of the first university, Bologna. In 1390, Dubrovnik also created the world's first permanent public health office to enforce its various public health rules.

Economic incentives helped motivate the trade-dependent city's innovations in public health and sanitation: "Sanitary measures in Dubrovnik were constantly improved because the city was forced to find a way to protect itself from diseases and at the same time retain the lucrative trade relations which formed its economic base," according to historians Tamara Alebić and Helena Marković. During the outbreak of 1347, the Dubrovnik writer and noble Nikola Ragnina (1494–1582) claimed that people first attempted to banish the plague with fire: "There was no cure and everyone was dying. When people saw that their physicians could not defend them, they decided to . . . purify the air with fire." The fires may have helped kill off some of the plague-carrying fleas, but were ultimately a failed experiment. So they tried something new.

Even a primitive understanding of how the illness spread proved sufficient for the people of Dubrovnik to attempt a radical and historic experiment in disease prevention. In 1374, Venice first put in place waiting periods for ship passengers to enter the city, but this was purely at the discretion of health bureaucrats, thus leading to irrational, selective enforcement. But in 1377, Dubrovnik's council implemented a much more logical system: all passengers on incoming ships and members of trade caravans arriving from infected areas were to wait for 30 days in the nearby town of Cavtat or the island of Mrkan before entering Dubrovnik's city walls. The quarantine period was soon expanded to 40 days (the word "quarantine" means "40 days")—a number likely reached as a result of experience, as the full course of the bubonic plague from contraction to death was typically around 37 days.

"Dubrovnik's administration arrived at the idea of quarantine as a result of its experience isolating leprosy victims to prevent spread of the disease," notes historian Ana Bakija-Konsuo. "Historical science has undoubtedly proved Dubrovnik's priority in the 'invention' of quarantine. Isolation, as a concept, had been applied even before 1377, as mentioned in the Statute of the City Dubrovnik, which was written in 1272 and . . . is the first mention of the isolation of the patients with leprosy." Dubrovnik's stone seaside quarantine shelters, sometimes considered the first plague hospitals in Europe, were called lazarettos after Lazarus, the patron saint of lepers. Today, the city's lazarettos serve as tourist attractions and concert venues.

Devastating plague outbreaks eventually forced Venice to implement a complete ban on anyone entering its walls, bringing trade and city life to a halt, but Dubrovnik's limited waiting periods let the republic keep its doors open to people and goods from abroad. "Hence, Dubrovnik implemented a method that was not only just and fair, but also very wise and successful, and it [eventually] prevailed around the world," according to historian Ante Milošević. Quarantine procedures remain the standard policy to this day when dealing with certain contagious diseases, such as Ebola.

The Black Death pandemic is sometimes viewed as the end of medieval civilization and the beginning of the Renaissance. Faced with a disease that would not become treatable until the advent of antibiotics in the 1940s, Dubrovnik certainly underwent a rebirth, recovering from the initial wave of deaths to become the first city to implement a coherent public health response to the bubonic plague. Dubrovnik's invention of quarantine represents not only perhaps the highest achievement of medieval medicine, but the emergence of one of humanity's oldest disease-prevention tools and a turning point in the history of public health. With its strong ideals of liberty and devotion to public health, Dubrovnik during its republican era has earned its place as a center of progress.

20

Benin City

SECURITY

Our next center of progress is Benin City, whose walls were once arguably the largest manmade structure on the planet. The wall network of Benin City was collectively four times longer than the Great Wall of China and consumed roughly 100 times more material to build than the Great Pyramid of Giza in Egypt, according to some estimates. Benin City was the capital of the Benin Empire (1180–1897), which was among the most highly developed states in sub-Saharan Africa before the European colonial period. Benin City was also known for its bronze artworks and a high degree of public order in its heyday. Prosperity requires physical safety from violence and property protection from theft or conquest, and the unprecedented scale of Benin

City's protective walls represented a significant achievement in security.

The walls of Benin City eventually fell to a military attack; however, the record-breaking structures successfully safeguarded the lives and property of those who lived within the city for centuries.

Today, Benin City is the capital and most populous city of Edo State in southern Nigeria, about 200 miles east of the economic and cultural hub of Lagos. Benin City is not to be confused with the country of Benin, which neighbors Nigeria to the west. A major urban center in coastal West Africa, Benin City is home to over 12 million people. Prominent local industries include rubber and oil production. Benin City is known for its festivals, for its rich dress culture, and for being the site of the royal palace of one of the world's oldest sustained monarchies—although today, the monarchy is largely ceremonial. The current *oba*, or traditional ruler or king, of the local people, was crowned in 2016 and is considered the 40th oba of Benin. His palace is a UNESCO World Heritage site.

The Benin Empire or kingdom—also sometimes called the Edo kingdom—originated sometime around the 10th century when the Edo people first settled in the rainforests of West Africa. The empire's traditional starting year is given as 1180. By the 15th century, the Benin Empire was an established regional power. The empire grew wealthier both by the conquest of neighboring territories and by robust trade with Europeans—initially the Portuguese and later the British. The word "Benin" comes from a mishearing of a word in the West African language Yoruba by Portuguese traders in the 15th or 16th century. The story goes that during a royal succession dispute, political pressures led an oba to renounce his office. As the oba gave up his title, in frustration, he publicly called the kingdom that he was surrendering the land of *Ibinu,* meaning "vexation" or "anger." In the form "Benin," the name stuck.

The capital of the Benin Empire was Benin City. If you could visit the city in its golden age, in the 17th century, you would observe a metropolis so orderly that theft was practically unthinkable within its walls. Lourenco Pinto, the Portuguese sea captain of a missionary ship, wrote in 1691:

> Great Benin [Benin City], where the king resides, is larger than Lisbon. All the streets run straight and as far as the eye can see. The houses are large, especially that of the king, which is richly decorated and has fine columns. The city is wealthy and industrious. It is so well governed that theft is unknown, and the people live in such security that they have no doors to their houses.

The city's security allowed the residents to be highly productive. Pinto also wrote, "The artisans have their places carefully allocated in the squares which are divided up in such a manner that in one square [I] counted altogether one hundred and twenty goldsmith's workshops, all working continuously."

Although Pinto wrote "goldsmiths," what he observed were almost certainly bronze workers. The city produced thousands of bronze plaques and sculptures created using a technique called low-wax casting. The city's historic artworks are widely considered to be among the best engravings made using that technique. Some of the bronzes depicted military exploits from the 16th-century period of rapid Benin Empire expansionism. Others represented trade and commerce, diplomacy, and dynastic history. However, most of the artworks were simply portraits of Benin's nobility garbed in elaborate ceremonial clothing.

Benin City's people also produced a great amount of cloth, which played a significant role in trade with European merchants. Other locally produced trade goods included pepper, palm oil, carved ivory, and beads made from cowrie shells and other materials. Benin City's people also sold slaves—often neighboring Africans captured in battle—to the Europeans. Distressingly, like almost all ancient societies, the Benin kingdom practiced slavery.

Despite its security achievements defending against outside threats, a modern person would not wish to live in ancient Benin City. Benin City's people practiced ritual human sacrifice with various rationales, including honoring a god of iron and petitioning the gods for profitable trade. The victims were often prisoners with criminal backgrounds. By the end of the 18th century, three or four human sacrifices occurred at the mouth of Benin City's river annually, ostensibly to ensure good trade with European merchants.

Among the city's most important imports were brass and copper ingots from the Europeans. The Benin Empire did not produce enough metal locally to supply Benin City's prolific engraving and sculpture industries fully. Many of the city's famed bronze artworks would not have been possible without the benefits of overseas trade. The Portuguese often sold bronze and copper to Benin City's people in the form of metal bracelets called *manillas*. By the 16th century, manillas and other metal objects (such as bronze pots and pans) were a standard trade currency used by Europeans in West Africa.

As trading grew more sophisticated, early factories or centers for the production of local goods like cloth sprang up along the main Benin City river. Always mindful of security, the Benin kingdom entered into various alliances to prevent piracy of trade goods.

Given the importance of trade to Benin City's success, it is fitting that one of the city's most beloved historical figures was a market woman. Her statue now graces a prominent place in Benin City. Emotan was a 15th-century merchant who, according to oral tradition, sold her wares at the point where her statue now stands. She founded the first childcare center in Benin City, opening up a nursery for the children of families patronizing Benin City's marketplace. She once warned a prince of Benin of a plot against his life and helped him regain the throne from his brother. The new king then rewarded her by appointing her to a high position charged with enforcing security in the marketplace. Emotan is now locally revered and deified as the "conscience of justice."

The city's wealth grew thanks to its thriving markets and international trade, as well as to the Benin kingdom's successful imperialism. As the city became richer, that wealth improved its infrastructure and many of its people's lives. "Houses are built alongside the streets in good order, the one close to the other," noted the 17th-century Dutch writer Olfert Dapper. "Adorned with gables and steps and roofs made of palm or banana leaves, or leaves from other trees . . . they are . . . usually broad with long galleries inside, especially so in the case of the houses of the nobility, and divided into many rooms which are separated by walls made of red clay, very well erected."

Dapper also noted that residents kept those walls "as shiny and smooth by washing and rubbing as any wall in Holland can be made with chalk, and they are like mirrors. The upper storeys are made of the same sort of clay. Moreover, every house is provided with a well for the supply of fresh water."

Benin City was also notably among the first urban centers to have a likeness of street lighting. Large metal lamps that burned palm oil, standing many feet high, were placed around the city.

The king's court was square and stood at the right side as one entered the city by its main gate. A wall like the one that encircled the city surrounded the court. The court housed various palaces, houses, and apartments for courtiers and boasted beautiful long, square galleries. Those galleries were "about as large as the Exchange at Amsterdam," according to Dapper. The largest gallery hosted many of the city's famed bronze carvings. Scenes engraved into bronze plaques stood supported by wooden pillars throughout the gallery.

However, Benin City's wall network would have been the city's most impressive sight. Radiocarbon dating of the walls' remains suggests that the Edo people built up Benin City's walls gradually over many years. Most of the construction likely occurred between 800 and 1500.

English author and journalist Fred Pearce wrote:

> The Benin [wall] network extend[s] for some 16,000 kilometres in all, in a mosaic of more than 500 interconnected settlement boundaries. They cover 6,500 square kilometres and were all dug by the Edo people. In all, they are four times longer than the Great Wall of China, and consumed a hundred times more

material than the Great Pyramid of Cheops. They took an estimated 150 million hours of digging to construct, and are perhaps the largest single archaeological phenomenon on the planet.

The walls extended, in other words, for about 10,000 miles and covered some 2,500 square miles. Since Pearce wrote those words, the official length of China's Great Wall—defined to include various distinct border walls built in all dynasties of Chinese history—has been updated to about 13,000 miles. However, that figure, calculated by China's State Administration of Cultural Heritage, has been called "misleading." It includes many isolated, disconnected walls defending various state boundaries within China, not just China's famed northern border wall. Estimates of the northern border wall's length vary from 1,500 to 5,000 miles, depending on exactly which wall sections are counted. In any case, Benin City's walls were certainly longer than China's storied northern border wall.

In places, the towers of Benin City's walls reached seven stories high. The walls also boasted guardhouses, ditches, moats, and garrison barracks. After guarding the city for centuries, the walls of Benin fell to British troops in 1897 during a "punitive expedition" motivated by British revenge for an earlier military strike by the Benin Empire. However, trade disputes also motivated the attack. Many of the royal palace's bronze artworks were captured in battle and are today displayed in the British Museum and various other museums.

After its walls fell, Benin City—and the Benin Empire—became part of the British Empire. Benin City then became part of Nigeria in 1960.

Safeguards against those who would steal or plunder have often proved indispensable in ensuring property rights. Although the walls of Benin City eventually fell, for centuries, the record-breaking security feature protected the city. Humans have created walls and other protective structures since they first switched from nomadism to permanent agricultural settlements. Many of the world's oldest stationary communities took the form of walled cities—such as our inaugural center of progress, Neolithic-era Jericho. However, the walls of Benin City surpassed all others in sheer scope. For the significant achievement in security of its record-shattering walls that stood for hundreds of years, Benin City is a center of progress.

21

Mainz

PRINTING PRESS

Our next center of progress is Mainz, the hometown of the inventor of the metal movable-type printing press, Johannes Gutenberg (c. 1399–1468), and the urban base from which that invention spread throughout Europe. Although he may have technically invented the printing press in Strasbourg, Gutenberg soon returned to Mainz, and it was in the latter city that he taught many others the art of printmaking. Political turmoil in the city soon caused a mass exodus of Gutenberg's apprentices. The printmakers spread out from Mainz to different corners of Europe, where they further disseminated printmaking knowledge. The Mainz printmaker diaspora helped increase the speed with which other parts of Europe adopted the printing press.

Today, Mainz is the capital and biggest city of Rhineland-Palatinate, a state in western Germany. The city is known for its wine production and beautiful reconstructed half-timbered houses and its market squares' medieval architecture. It is also a carnival stronghold: particularly in the days leading up to Ash Wednesday, the city brims with parades and music. Located on the Rhine River's banks, Mainz also houses a beautiful cathedral dating back to 975, as well as the Gutenberg Museum. The museum, devoted to the history of printing, was founded in 1900 and contains two original 15th-century Gutenberg Bibles. The city's industries are varied and include chemical and pharmaceutical products, electronics, precision instruments, machinery, glassware, and musical instruments. Appropriately, given its history, Mainz also remains an important media center, with publishing houses as well as radio and television studios. The city also honors its most famous resident with a festival in his honor each summer, called Johannisnacht.

The Romans founded Mainz at the site of a preexisting Celtic settlement in the first century BCE, establishing it as a military fortress, or *castrum*, on their empire's northern frontier. They named the outpost Moguntiacum, after the local Celtic deity Mogo or Mogons, likely a god of battle. The Latin name Moguntiacum eventually evolved into the German name Mainz, which the city bears to this day. The Romans introduced vineyards to the area, and winemaking remains a key local industry. The Roman conquerors also brought the Latin writing system with them—a writing system with a limited alphabet that, as we shall see, likely bolstered the eventual success of the printing press. Mainz also served as a provincial capital of the territory that the Romans called Germania Superior.

Mainz again rose to political prominence in the ninth century when it began to serve as the capital of the Electorate of Mainz in the Holy Roman Empire. The Holy Roman Empire was a political institution that, for centuries, united different constituent territories or kingdoms in central and western Europe into something more akin to a confederation than a true empire. Constituent principalities had their own rulers and enjoyed relative independence. The *Encyclopedia Britannica* calls the Holy Roman Empire, "along with the papacy, the most important institution of western Europe" during the Middle Ages. And the Electorate of Mainz is widely regarded as having been one of the most prestigious and influential states within the Holy Roman Empire. Mainz was the seat of the archbishop-elector of Mainz, the primate of Germany. (The primate of Germany was a historical title given to the most powerful bishop in the German-speaking areas of Europe.) This archbishop acted as the archchancellor of Germany, one of the constituent kingdoms of the Holy Roman Empire, and was second in power only to the Holy Roman emperor.

In other words, by the time Gutenberg was born in a house situated on a corner in Christofsstraße in Mainz, in the late 14th century—the spot is now marked by a commemorative plaque—the city was a well-established center of political importance. But the city was deeply unstable, wracked by internal disputes and economic turmoil.

Tensions between the city's patricians or nobility and the fast-growing merchant class were palpable throughout Mainz. In 1332, to quell a brief civil war, the archbishop of Mainz granted the guilds representing merchants and craftsmen equal representation on the city council alongside the old nobility.

But by the early 1400s, Mainz was home to more merchants and guild members than patricians, and conflict between the groups was again frequent. In 1411, an uprising of merchants protesting special tax and customs privileges reserved to the nobility occurred. The protesting rioters set the homes of several patricians on fire. Afraid for their lives, 117 patricians fled Mainz amid the turmoil, including the family of the young Gutenberg. The family soon returned to Mainz, but the city only grew more troubled. Periodically fleeing Mainz was a recurring theme in the life of Gutenberg and many other Mainz residents.

The extent of instability within the city was so disruptive that it contributed to shortages of basic goods. In 1413, food became scarce throughout Mainz. As the city's people starved, mass hunger riots broke out, resulting in much violence and property destruction. The riots prompted the Gutenberg family, and many others, to flee Mainz once again.

Gutenberg returned to Mainz, always drawn back to his hometown despite its problems. As he entered adulthood, Gutenberg found himself not quite fitting into either warring faction within the city. Many people hated him for his patrician status, which Gutenberg inherited through his father. Still, the city did not grant him the special legal privileges reserved to most patricians because his mother was a commoner by birth. Understanding his precarious position and attempting to safeguard his economic future, Gutenberg took up the metalworking trade.

By 1428, the city of Mainz teetered on the verge of bankruptcy, and the power dispute between the patricians and guild members entered a new phase in which the guilds seized power.

As the city reeled from internal violence, tribal prejudices, and a crashing economy, many people understandably fled Mainz. Gutenberg was probably one of them, and in any case, he was living in Strasbourg by 1434.

In Strasbourg, Gutenberg transcended his era's tribalism and strategically befriended both patricians and guildsmen, although he did not join the metalworking guild. Leveraging his connections with local officials, Gutenberg successfully pressured a visiting official from Mainz to pay him a debt that the city of Mainz owed his family and likely used the capital to bolster his metalworking business. There is also evidence that Gutenberg briefly dabbled in the region's prominent wine trade. While he lived in Strasbourg, Gutenberg probably developed the metal movable-type printing press.

It must be noted that the Chinese invented woodblock printing many centuries earlier. An inventor in Hangzhou—which you may recall as another center of progress featured in this book—even devised movable type, as early as the 11th century. However, several factors prevented movable type from seeing the level of widespread adoption in China that the technology achieved in Europe. Those factors ranged from the cultural importance that many Chinese placed on handwritten calligraphy to the sheer number of characters in the Chinese writing system. There are thousands of different Chinese characters. Contrastingly, German uses a limited Roman alphabet, which made printing the language more practical.

In 1448, Gutenberg went home to Mainz. As Alexander Hammond wrote in his profile of Gutenberg for HumanProgress .org, "With the help of a loan from his brother-in-law, Arnold

Gelthus, he was able to build an operating printing press in 1450." Initially, Gutenberg marketed his innovation as a way to allow monks to reproduce religious texts at a much faster rate. He maintained two presses: one for the Bible and one for commercial texts. By 1455, he printed the first 180 copies of the "Gutenberg Bible." The printing press proved an initial success, allowing Gutenberg to take on apprentices and locally disseminate printmaking knowledge. Unfortunately, a lawsuit by an investor left him near bankruptcy.

The city of Mainz continued to deteriorate in a downward spiral, as a period of economic decline culminated in war between two rival archbishops. The Mainz Diocesan Feud, also known as the Baden-Palatine War, took place in 1461–1462. The combatants fought over the throne of the Electorate of Mainz. Following a close election to become the new archbishop of Mainz, both Diether von Isenburg (the victor by a small margin) and Adolph von Nassau declared themselves the rightful archbishop. With the help of their respective political allies, Diether and Adolph went to war. Diether had made enemies of both the pope and the Holy Roman emperor Frederick III, and the latter two thus backed Adolph's claim. Many people in the city of Mainz, including the city council, continued to support Diether, who refused to vacate the city or his archbishop's throne.

Adolph and his troops sacked the city, and eventually Adolph prevailed in seizing control. In 1465, Archbishop Adolph recognized Gutenberg's contributions to human progress by granting him a court position and a large annual stipend, allowing Gutenberg to live the rest of his days in relative peace and comfort in Mainz, where he is buried.

If you could visit Mainz during the city's sacking, you would have borne witness to a scene of terrifying violence and destruction. You also would have seen an exodus of the city's fleeing people. Some of those people carried with them knowledge that would change history.

Almost all of the centers of progress featured in this book have contributed to progress during ages of relative peace and prosperity, but in Mainz, that was not the case. Instead, the city's instability became a catalyst for change. The city's economic and political turmoil drove many of Mainz's craftsmen into exile, including Gutenberg's printing apprentices, thus spreading the knowledge of the art of printing throughout the European continent with incredible speed.

According to some estimates, by the 1470s, a mere decade later, every major European city had printing companies (see the map on the next page), and by the 1500s, about 4 million books had been printed and sold. The ability to reproduce the written word so quickly brought the spread of new ideas. Ranging from the Protestant Reformation to the later Enlightenment and the rise of new forms of government, several massive societal transformations came about largely because of the possibilities presented by the printed word.

"Every time the cost of media declines rapidly, you enable more people to speak out, and you have a greater diversity of voices," according to American historian Bill Kovarik, explaining that this affects the distribution of power in society and sparks social change. Today, the digital revolution has further lowered the cost of disseminating ideas and knowledge, continuing the revolution in communications that began with Gutenberg's printing shop in Mainz.

Fifteenth-century printing towns of incunabula

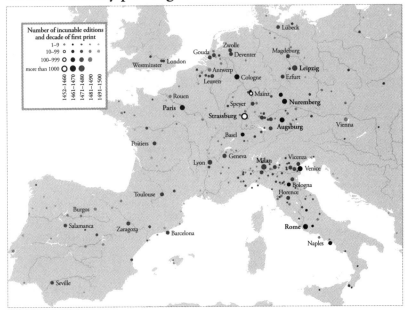

Plagued by violence and economic problems, Mainz during the 15th century was an unlikely site of progress. But the invention that spread with incredible speed thanks to the diaspora of printmakers fleeing the city was pivotal to the future of human progress. The printing press ultimately helped erode the power of the guilds and the nobility, the very same warring factions that caused so much turmoil in Mainz. By democratizing the spread of information, the printing press enabled the proliferation of everything from scientific and medical texts to philosophical and political treatises. For those reasons, Mainz, the city responsible for Europe's rapid adoption of the printing press, is a center of progress.

22

Seville

NAVIGATION

The next center of progress is Seville during Europe's Age of Exploration, from the mid-15th to the mid-16th century, when the city was a major trade port at the forefront of progress in maritime navigation. In 1519, a five-ship expedition departed Seville on a quest to sail around the world. In 1522, only a single ship from that expedition returned, the galleon *Victoria*. And victorious she was, having sailed 42,000 miles to successfully circumnavigate the globe—a milestone in the history of navigation.

Today, Seville is the capital of, as well as the most populous and richest city in, Andalusia, and its harbor remains busy as Spain's only commercial river port. The port handles exports such as wine, fruit (including oranges, which famously grow throughout and perfume the city of Seville), olives, and minerals. The port also handles imports, including oil and coal. Shipbuilding is a major part of the city's economy, alongside the services industry and tourism. The city is known as the world capital of flamenco, and throughout the city there are frequent performances of that dance style, which is likely a fusion of Asian and European dance forms brought about by a wave of immigration from northwestern India to Andalusia between the 9th and 14th centuries. The city is also known among tourists for its bullfights and its religiosity, with many believers thronging to the city during its *Semana Santa* (Easter Holy Week) festivities. The city is also the setting of several famous operas, including the *Barber of Seville*.

Seville's architectural wonders have been featured as backdrops in famous movies and television series, including *Star Wars* and *Game of Thrones*. Although the city contains notable modernist buildings, such as the world's largest timber-framed structure, the distinctive *Las Setas* (the Mushrooms), Seville remains best known for its historic architecture. The city's Old Town contains no less than three UNESCO World Heritage sites. One is the Mudéjar-style Alcázar royal palace, which was largely built by Castilians in the 14th century on the site of an earlier Abbasid dynasty–era (1023–1091) fortress, incorporating some of the original structures, to house King Peter the Cruel (1334–1369). To this day, Spain's royal family continues to occupy the Alcázar when visiting Seville, making it Europe's oldest royal palace still in use.

Another World Heritage site is Seville Cathedral, which took more than a century to build. Completed in 1507, it presented an extravagant sight during Seville's golden age of trade, just as it does today. It remains the world's largest Gothic-style church, as well as the fourth-largest church of any kind. It is said the original construction committee was tasked to create something "so beautiful and so magnificent that those who see it will think we are mad." The surrounding orange trees delight the church's visitors with Seville's trademark scent.

The city's final World Heritage site is the General Archive of the Indies. It was commissioned in 1572 by King Philip the Prudent (1527–1598), who ruled during the peak of the Spanish Empire, to serve as the Merchants' Exchange House (*Casa Lonja de Mercaderes*) for Seville's tradespeople to conduct business related to their New World voyages. Throughout its history, different portions of the enormous Renaissance building have variously served such diverse functions as a painting academy, a grain storehouse, and a shelter for orphans and widows. As its current name suggests, the General Archive of the Indies now serves as a repository of archival documents illustrating the history of the Spanish Empire and its transatlantic trade.

According to mythology, Seville's founder was none other than the famous demigod hero of classical literature, Hercules. A garden square called the *Alameda de Hércules* (Hercules Mall), built in 1574, greets visitors to this day with a towering statue of the hero. More precisely, the city's mythical founder was the Phoenician god Melqart, who later became identified with Hercules. The oldest part of Seville was likely constructed around the eighth century BCE, on an island in the Guadalquivir River (derived from the Arabic *al-wādī l-kabīr*, meaning "the great river"). Seville was multicultural from its inception, defined by a mingling of the Tartessians, an indigenous Iberian people, and Phoenicians, who were lured by the city's potential as a trade port.

Seville's geography perhaps destined it to become a major port. The city marks the point on the 408-mile-long Guadalquivir river, beyond which ships are unable to travel farther inland. As Spain's only major navigable river, the Guadalquivir has been used to transport goods since at least the eighth century BCE, when the ancient Phoenicians moved precious metals that were mined in Spain by boat, carrying them out to sea and delivering the cargo to the world's first major port at Byblos, in what is today Lebanon, as well as to the Assyrians. The river was not only the main artery of trade traffic in and out of Andalusia, but it provided access to the Atlantic, which became critical for exploration of the New World and, ultimately, the achievement of circumnavigating the globe.

Over the years, Seville was ruled by Carthaginians, Romans (whose city walls remain partially intact), Visigoths, Moors, and Castilians. The city was always a prominent trade gateway and was progressively diversified by a constant influx of goods and people from different cultures. But it was during Spain's golden age, at the height of the Spanish Empire's transatlantic trade in the 16th century, that Seville grew to be one of the largest cities in western Europe.

If you could visit Seville during its glory days, you would enter an intoxicating metropolis with eclectic architecture epitomizing centuries of cultural intermingling. As a Spanish rhyme goes, *Quien no ha visto Sevilla, no ha visto maravilla* (He who has not seen Seville, has not seen wonder). Walking among the crowds along the city's intricately tiled pathways and mosaic plazas, you would have seen a thriving cosmopolitan commercial center that housed merchants from across the continent, and you would have heard not only Spanish, but English, Flemish, and Italian, among other languages. Although Islam was outlawed in 1502, there remained a significant Moorish, formerly Muslim minority, known as Moriscos, some of whom continued to practice Islam in secret. Enslaved Africans also would have been present. The streets would be abuzz with talk of the latest groundbreaking maritime expeditions, as Europe's great powers competed for mastery of oceanic trade avenues and raced to be the first to discover promising sea routes and uncharted lands.

In 1503, Spain granted Seville exclusive trading rights with the New World, and the city prospered. But as British historian Richard Cavendish has noted, "The idea that such a web of human activity could be controlled by a bureaucracy proved hopelessly unrealistic and for all the cascade of silver, Spain remained a poor country." The weight that Seville's monopoly and other policies limiting economic freedom put on the Spanish economy contributed to the government's financial troubles, including nine eventual bankruptcies of the Spanish monarchy (in 1557, 1575, 1596, 1607, 1627, 1647, 1652, 1662, and 1666). Government-backed privileges enjoyed by the elite in areas ranging from trade to land management, monetary inflation from an influx of New World silver, and high government war spending were some of the factors that stymied economic development. Seville's golden age was short-lived, ending when the crown transferred control of New World trade to Cádiz in 1717.

Among 16th-century Seville's crowds, you might have glimpsed the renowned novelist Miguel de Cervantes (1547–1616), who studied at the Jesuit college in Seville in the 1560s and returned to the city in 1588 for a few years. Seville featured in several of his works, for example, by providing the setting for his short story about the city's organized crime scene, *Rinconete y Cortadillo*. In one poem, Cervantes characterized the city this way: "O great Seville! Like Rome triumphant in spirit and nobility." In Cervantes's magnum opus, *Don Quixote*, the eponymous central character of that groundbreaking novel (first published in 1605) receives an invitation to visit Seville because "it was just the place to find adventure, for in every street and on every street corner there were more adventures than in any other place."

And a sense of adventure surely must have filled the air on the fateful day when an expedition departed Seville's harbor on a quest to circumnavigate the globe. The achievement came at a cost: the expedition set out with some 260 people, but only 18 of them returned to Seville after circumnavigation. True to Seville's multicultural reputation, the survivors who completed the voyage represented a number of nationalities. There were three Galicians, three Castilians, two Greeks, a Venetian supernumerary, a Genoese chief steward, a Portuguese mariner, a German gunner, and six Basque crew members, including the expedition's ultimate captain, Juan Sebastián Elcano (c. 1486–1526). The Venetian, Antonio Pigafetta (c. 1491–c. 1531), kept precise journals chronicling the voyage, which many scholars consider the most reliable account of the expedition. Absent from the return was the Portuguese explorer Ferdinand Magellan (1480–1521), who had planned the Spanish expedition, but was killed en route in the Philippines. (More details can be found later in this chapter.)

Europe's Age of Discovery saw competition between many countries, but Portugal and Spain led the way. At first, Portugal dominated, discovering and claiming the Atlantic archipelagoes of Madeira and the Azores in 1419 and 1427, respectively, and finding a game-changing sea route to India in 1498, around Africa's Cape of Good Hope. The Portuguese focus on navigation even resulted in the unusual royal nickname of Prince Henry the Navigator (1394–1460). In 1501, the Florentine Amerigo Vespucci (1451–1512), while on a Portuguese expedition looking for another maritime route to Asia, discovered what he called the New World—and from his name we get the word "America." Advances in shipbuilding—including the development of a stabler, faster, and more maneuverable kind of ship called the galleon toward the beginning of the 16th century—further accelerated progress in navigation.

Although the governments sponsoring the expeditions may have been rivals, major voyages tended to be enterprises of multinational cooperation, with crew members hailing from many countries—including Spaniards serving on Portuguese expeditions and vice versa, as the Spanish and Portuguese crowns competed to hire the best talent. Spain began to challenge Portugal's supremacy of the seas in part thanks to its openness to expertise from abroad. It was Spain that financed the Genoese explorer Christopher Columbus (1451–1506) for his famous 1492 voyage to the Americas, which he mistook for the East Indies. (The West Indies owe their name to that mistake, and "Columbia," derived from Columbus's name, remains a poetic term for America.) The Florence-born Vespucci died a Spanish citizen in Seville, with the title of Spain's chief navigator, in 1512. The Spanish explorer Vasco Núñez de Balboa (1475–1519) became the first European to cross the Americas to the Pacific Ocean in 1513, and in 1516, Juan Díaz de Solís (1470–1516), who may have been born in either Seville or Lisbon, became the first European to reach Uruguay, while on a Spanish expedition.

Magellan dreamed of finding a direct trade route to the Spice Islands, in what is today Indonesia, that avoided having to go around Africa with its many rocky outcrops. The treacherous Cape of Good Hope route had become known as a ship graveyard. After repeated failed attempts to solicit funding for his voyage from Portugal's monarch, Magellan went to Seville in 1517 to try his luck with the Spanish crown. Supportive of Magellan's vision, but heavily in debt, Spain's teenage king—the future Holy Roman emperor Charles V (1500–1558)—was unable to fully fund the voyage. The private sector stepped in to make Magellan's expedition possible. Cristóbal de Haro (d. 1541), a Burgos-born financier and merchant connected to the Fuggers, a prominent German banker family, provided the critical remaining funds that were needed for the voyage, as well as goods for the crew to barter.

In 1519, Magellan departed Seville with a five-ship fleet consisting of the flagship *Trinidad*, the *San Antonio*, the carrack *Concepción*, the *Santiago*, and the *Victoria*. The *Victoria* was born as the *Santa Maria* in the shipyards of Ondarroa in Spain's north and was used for trade between Castile and England before the crown purchased the vessel in 1518. Magellan renamed her after his favorite chapel in Seville, the Santa María de la Victoria.

After a long journey across the Atlantic and travels along the South American coast in search of a route to the Pacific, the *Santiago* was wrecked in an Argentinian river in 1520 during a storm. Later that year, the expedition discovered a navigable sea route to cross the Americas to the Pacific through Chile, which was later named the Strait of Magellan. Until the Panama Canal's completion in 1914, the strait provided the only relatively safe maritime path between the Atlantic and Pacific Oceans. At the strait, the *San Antonio* deserted the expedition and returned to Spain. To justify their desertion, the crew described Magellan as a psychopath. His reputation in Seville suffered and his wife and child were sentenced to house arrest. It was only after Pigafetta disseminated his account of the voyage that Magellan's reputation recovered. To this day, opinions of Magellan vary wildly.

After crossing the strait, Magellan named the body of water beyond it the Pacific Ocean because its waters were peaceful when he entered it. Unaware of the ocean's vastness, the explorers expected to cross it in a few days, but it took months before they made landfall. By that point, the crew had exhausted their food supply and were reduced to devouring ship rats and sawdust. The majority developed scurvy, a condition caused by a deficiency of vitamin C, and many died of malnutrition.

But their troubles did not end when they finally reached Guam and the Philippines. An enslaved crew member and native speaker of Malay, Enrique of Malacca (1495–after 1522), conversed successfully with the locals, proving they had indeed reached Asia. He may have been the first person to circumnavigate the world, if one assumes the voyage brought him back to the land of his birth. (His origins are unclear—he may have been Malaysian, Indonesian, or born elsewhere.) The expedition soon became embroiled in conflict—the Battle of Mactan. Magellan led a contingent of his crew to fight for a local ruler, Humabon of Cebu, against the warriors of Lapulapu, who was the chieftain of Mactan, an island located about one mile east of Cebu. Magellan's crew was better armed; however, Lapulapu's men outnumbered them, and they killed Magellan with a poisoned arrow. Today, Indonesians celebrate a holiday honoring Lapulapu for defeating the foreign force, and a prominent shrine in Mactan features a statue of Lapulapu and a mural painting of Magellan and Lapulapu in combat. According to Pigafetta, after the crew refused to free Enrique upon Magellan's death, as specified in the latter's will, Enrique successfully conspired with Humabon to arrange their extermination and his own freedom. Humabon invited part of the expedition, including the crew astrologer, San Martin of Seville, to a feast and had them massacred. The expedition's survivors scuttled (deliberately sank) the *Concepción* in 1521 because they no longer had enough men to crew three ships, and the *Trinidad* later broke down in the Spice Islands.

When the *Victoria* finally docked in Seville's harbor three years after her departure, loaded with spices, she fired off salutes with the expedition's remaining gunpowder. Pale and emaciated, the crew slowly disembarked, forever scarred by the memory of mutinies, disease, starvation, war, and storms at sea. Their leader Elcano called them "the skinniest men there ever were." The cheering multitudes of Sevillians that greeted them handed out candles and applauded as the expedition members walked shakily and wordlessly to their ship's namesake, the shrine of Santa María de la Victoria, to give thanks for their survival. Today, a slab in the cathedral honors them. Throughout their ordeal, the first circumnavigators of the Earth contributed profoundly to humanity's navigational understanding: they found the Strait of Magellan, learned of the immensity of the Pacific Ocean, and confirmed that the world was round. The spices they transported to Seville were valuable, but the greatest treasure they brought home was their hard-won navigational knowledge.

Bursting with commerce and adventure seekers from across Europe, 16th-century Seville gave humanity the first expedition to circle the globe, which historian Laurence Bergreen called "the greatest sea voyage in the Age of Discovery." That age ushered in a new phase of sea-based globalization that expanded humanity's horizons, allowing the mapping of the world. As distant civilizations came into contact, brutal conflict often occurred, including the transatlantic slave trade and colonial power struggles. But the global interconnectedness enabled by mankind's newfound navigational expertise ultimately helped create modern society, with far-reaching exchange of scientific knowledge and the prosperity generated by worldwide trade. Financed by both kings and merchants, the expeditions that departed Seville's harbor undoubtedly changed the world. It is for those reasons that Seville has found its way to these pages as a center of progress.

23
Amsterdam
OPENNESS

Our next center of progress is Amsterdam between the founding of the Dutch Republic in 1581 and the French military invasion of 1672. During the Dutch Golden Age, Amsterdam was an early center of globalization, exemplifying openness to foreign ideas, people, and goods. In the 17th century, the Dutch opened up a global trade network with the Far East and gained an increasing share of world trade. The city was also remarkably tolerant with respect to religious and intellectual freedoms. Controversial philosophers and religious refugees alike found a safe haven in the city. Amsterdam served as the headquarters of the world's first multinational corporation, the Dutch East India Company, which was founded in 1602. Amsterdam can take credit for housing the first modern stock exchange, which has traded continuously since the early 17th century, and is commonly considered the world's oldest true securities market. As trade and financial innovations enriched the city, Amsterdam also became a global leader in science and art.

Today, the port city of Amsterdam serves as the Netherlands' capital but not the country's government seat, which is in The Hague. It is also the Netherlands' most populous city. The city is nicknamed the Venice of the North, due to its many canals dating to the 17th century, which are recognized as a UNESCO World Heritage site. The city's fortifications, built between 1883 and 1920 and known as the Defense Line of Amsterdam, compose a separate UNESCO World Heritage site. The city is famous for its nightlife, and many historic sites and museums, such as the Van Gogh Museum. The Dutch royal palace is also in Amsterdam, although the current royal family does not use it as their primary residence. The city remains the Netherlands' commercial center, as well as one of Europe's top financial centers. Amsterdam is also one of the world's most multicultural cities, with at least 177 different nationalities represented among its residents. That extraordinary degree of multiculturalism is a long-standing part of the city's fabric.

Amsterdam gets its name from the city's origins as a fishing village that sprung up in the 12th century in a flat, low-lying area by a dam in the river Amstel. Parts of Amsterdam are below sea level, on land the Dutch successfully reclaimed from lakes, marshes, and the North Sea. As Swedish writer Joakim Book put it: "For centuries, the people living along the Atlantic coasts have carved off and dammed areas when the tide went out, gradually drying saltmarshes and expanding land. . . . Today upwards of one-third of this prosperous northern European nation's territory lies below sea level . . . the Dutch have 'one of the most sophisticated anti-flood systems in place anywhere in the world.'" Even today, the Dutch can claim some of the world's best hydraulic engineers, and the American Society of Civil Engineers placed the country's water protection systems on its list of the Seven Wonders of the Modern World.

Historically, Amsterdam was a center not only of ingenuity but also of tolerance and openness. During Europe's wars of religion, Amsterdam was a refuge for Protestants of varying kinds, including French Huguenots. The city prided itself on granting freedom of conscience (*geweten*) to all residents, which was in keeping with the beliefs of the city's reigning Calvinist faction. The city's tolerance did not meet modern standards, of course. Public displays of Catholicism were illegal, and Catholic churches had to meet highly restrictive criteria and remain hidden from public view. But in an era when religious intolerance could be lethal and even different, and Protestant denominations often bitterly opposed one another, Amsterdam took a relatively open-minded approach. Amsterdammers embraced and courted skilled and wealthy foreigners of various creeds, seeing internationally well-connected intellectuals and merchants as valuable members of society. They did that at a time when many other European countries were becoming more insular and religiously intolerant.

Amsterdam became a bustling cosmopolitan metropolis as the population doubled to about 50,000 between 1570 and 1600. By the year 1600, one-third of Amsterdammers were foreign born. Amsterdam was also the center of the Dutch Jewish community. The revolt against the Spanish that led to the

Dutch Republic's founding and kicked off the country's golden age also led to an influx of Iberian Jews seeking greater religious freedom. Amsterdam soon welcomed Jewish refugees from the Thirty Years' War (1618–1648) and the Cossack-Polish War (1648–1657). To this day, one of the city's nicknames is *Mokum*, a Yiddish word meaning "place" or "safe haven." (Centuries later, during World War II, Anne Frank and her family famously hid from Nazi persecution in an Amsterdam canal house built in the 17th century; her family moved to Amsterdam in 1934 when they fled Frankfurt.)

Amsterdam's tolerance helped the city grow. By the 1660s, near the end of the Dutch Golden Age, the city's population swelled to 200,000. That is about the same population as Madison, Wisconsin, today. (For perspective, the world's biggest city at the time was probably Constantinople or Beijing, both with more than 700,000 people.)

Amsterdam was central to the Dutch Golden Age, when the Netherlands rose from a small, obscure nation on the North Sea to become one of the world's most influential countries. In fact, Amsterdam has been called the "capital of the Golden Age." Thus, it can be said that the Netherlands' rise stemmed from a rapid and continuous economic expansion centered in Amsterdam.

Among the chief Dutch exports were cheese and fish from the North Sea, such as herring. In 1602, various rival Dutch trading companies joined forces to form the world's first multi-national corporation, headquartered in Amsterdam. The Dutch East India Company facilitated trade with Mughal India during the latter's period of proto-industrialization. The company imported goods such as textiles and silks, provided shipping, and diversified into various other commercial activities. For its complexity, it has been labeled a proto-conglomerate. The Dutch East India Company has also been called the prototype or forerunner of the modern corporation. The megacorporation was both a transcontinental employer and trailblazer of foreign direct investment. The company's formation was arguably a key episode in the dawn of modern capitalism. It must be noted that the company was, appallingly, also tied to the Dutch slave trade and colonial expansionism. Slavery was still common in many societies at the time, and the Dutch colonists were no exception.

Amsterdam's openness to trade extended beyond traditional goods and services; Amsterdammers also traded stocks. It is true that Bruges was home to the first bourse, where traveling Italian bankers traded securities in the Van der Buerse family's inn (from which the word "bourse" is derived) in the early 15th century. However, most scholars agree that Amsterdam can take credit for housing the first stock exchange in the modern sense. The Dutch East India Company established the Amsterdam Stock Exchange in 1602 and became not only the first modern corporation, but also the first corporation in the world to be listed on a stock exchange.

Long-distance trade by ship was a risky undertaking, with goods traveling from Asia to Europe susceptible to loss in shipwrecks or theft by pirates. The stock exchange allowed the company to spread out the risks—as well as the dividends—of international trade among a broad pool of investors. When a voyage ended in a shipwreck, no single entity had to bear the full cost of the loss. When an expedition was successful, many investors benefited. Shareholders soon gained the ability to transfer their shares to third parties, and by the mid-17th century, the flourishing stock exchange inspired the formation of "trading clubs" around Amsterdam. Those clubs met in coffee shops or inns throughout the city to discuss transactions and cultivated a growing community of traders. The Netherlands was also, unfortunately, home to the first great speculative financial bubble, when tulip bulb contract prices in the futures market soared to unprecedented highs before collapsing in 1637.

The French historian Fernand Braudel disputed the broadly held view that Amsterdam housed the "first modern stock exchange," but conceded that Amsterdam's exchange had historical significance: "What was new in Amsterdam was the volume, the fluidity of the market and publicity it received, and the speculative freedom of transactions." In sum, the amount of trading activity that occurred in the Amsterdam Stock Exchange was unprecedented.

Amsterdam grew increasingly prosperous, thanks to its role as a financial center and a key player in international trade. As the Dutch Republic became one of the world's richest countries, the Dutch poured funds into science and art. During that era, the Dutch invented microbiology, discovered Saturn's moon Titan, and invented the pendulum clock. The Dutch Golden Age also produced some of history's most beloved painters, such as Rembrandt (1606–1669), who worked in Amsterdam; and Vermeer (1632–1675), who lived in Delft but who received artistic funding from Amsterdammers, including the silk trader and art collector Hendrick Sorgh (1666–1720).

Amsterdam's renowned tolerance attracted cutting-edge thinkers like the French philosopher René Descartes (1596–1650) and the English "father of liberalism" John Locke (1632–1704) to take refuge there for a time. The city's atmosphere also gave native Amsterdammers like the philosopher Baruch Spinoza (1632–1677) the intellectual freedom to explore their ideas. Amsterdam was willing to print many controversial books that other European cities would not, incentivizing various intellectuals abroad—such as the English political philosopher Thomas Hobbes (1588–1679)—to arrange to have their books printed in the Dutch city.

The Dutch Golden Age came to an abrupt end in 1672, which is commonly called the *Rampjaar* (Disaster Year), when the Franco-Dutch War broke out. The French troops and their allies nearly overran the Netherlands and caused great destruction. The Dutch only managed to stop Louis XIV's advance by intentionally flooding their own country. The Dutch had developed an ingenious defensive system called the Dutch Water Line, which could quickly flood the nation and transform the Netherlands into something close to a set of islands. The Dutch had used intentional flooding as a military tactic dating to the Dutch war of independence (1568–1648), but the Dutch Water Line took the concept to a new level. The Dutch deliberately flooded their country with a layer of water that was too deep to allow an invading army to advance by foot but too shallow for boats to traverse. The flood brought movement across the Netherlands to an effective standstill and stopped the French invasion.

Perhaps no city better exemplifies the benefits of social openness and early globalization than Dutch Golden Age–era Amsterdam. By embracing foreign peoples, goods, and ideas, what began as a small fishing town became a prosperous global capital of philosophy, science, and art. Far-ranging trade, new corporate structures, innovations in finance and engineering, and acceptance of intellectual and religious refugees all helped make Amsterdam successful. For its myriad groundbreaking achievements and the underlying attitude of openness that helped enable them, 17th-century Amsterdam is deservedly a center of progress.

24

Agra

ARCHITECTURE

The next center of progress is Agra during the city's golden age at the time of the Mughal Empire (1526–1857). In Agra, different cultures converged to create what many believe is humanity's greatest architectural achievement: the Taj Mahal (constructed 1631–1653).

Located on a broad plain on the banks of the Yamuna River in the Indian state of Uttar Pradesh, Agra is home to roughly 1.6 million people. The city is known for its leather goods, hand-woven carpets, stone handicrafts, and distinct red sandstone. It is also known for its Mughlai cuisine, which has evolved considerably from the days when Mughal emperors dined on food flecked with silver. As a major road and rail junction, as well as a prong of India's "golden triangle" tourist circuit, Agra is a transportation hub. Tourism is a major factor in Agra's economy, and the city contains two UNESCO World Heritage sites: the Agra Fort and the Taj Mahal.

The area where Agra now stands has attracted notice since ancient times. Agra is referenced in the ancient Sanskrit epic poem the *Mahabharata*, which mentions "the forest of Agravana." But it was the famed scholar Claudius Ptolemy (c. 100–c. 170), who lived in Alexandria nearly 4,000 miles away, who provided history's first recorded use of the name "Agra." "It is easy to recognize the Yamuna, the river, which after passing Delhi, Mathura, Agra, and other places, joins the Ganges," as Ptolemy noted in his work *Geographia* (*The Geography*), published in 150.

Despite these ancient roots, according to tradition, Agra was founded in the year 1504, when Sultan Sikandar Lodi made it the capital from which he and later his son, Sultan Ibrahim Lodi, ruled over the Delhi sultanate.

But Agra truly rose to prominence under the Mughal Empire, founded by the Uzbek-born chieftain Babur (1483–1530) in 1526, who conquered Agra and took the younger Lodi's throne. He had the Ram Bagh (Garden of Relaxation) laid out on the banks of the river Yamuna, where it remains as the oldest extant Mughal garden. Babur's daughter-in-law, Empress Bega Begum (c. 1511–1582), began the dynasty's tradition of palatial tombs on the Indian subcontinent in 1558 when she commissioned an elaborate final resting place for her husband, Babur's son, the second Mughal emperor Humayun (1508–1556). Created by architects from Persia and representing the first garden tomb in India, this impressive structure in Delhi would soon be dwarfed by the tombs of Agra.

The empire greatly expanded under Humayun's son, the third Mughal emperor, Akbar the Great (1542–1605). Akbar focused on both territorial and commercial expansion, conquering land and strengthening trade ties with neighboring realms. Agra's population swelled under Akbar, reaching as many as 800,000 people.

Akbar redesigned and raised the towering ramparts of the Agra Fort and commissioned the 15-story-tall Buland Darwaza (Door of Victory) just outside of Agra, which remains the highest gateway in the world. Akbar was, for his era, unusually tolerant of other religions. He repealed the customary tax on non-Muslims (the *jizyah*) and abolished the death penalty for deconverting from Islam back to Hinduism. He created a religious institution known as the *Ibādat Khāna* (House of Worship), which encouraged interfaith philosophical and theological debates.

Akbar also personally engaged in a radical experiment in religious syncretism, promulgating what some historians describe as a spiritual training program and others call a new religion. The movement, called Din-i Ilahi (Divine Faith), attempted to reconcile and merge Islam, Hinduism, and Zoroastrianism and incorporated elements from Christianity, Jainism, and Buddhism. Akbar sought to promote what he saw as the best aspects of these different faiths—such as Hinduism's encouragement of vegetarianism and Islam's central tenet of *Tawhid*, or monotheism. Many of his Muslim contemporaries considered the emperor a heretic (as do many Muslims today), but Akbar's unusual views helped increase his popularity among his many Hindu subjects.

Akbar took up various native customs, participated in Diwali and other local festivals, and showed an enthusiasm for Sanskrit literature, which he had translated. His son Jahangir and grandson Shah Jahan also would not eat beef in deference to Hindu beliefs. Multiculturalism continued to define the Mughal Empire for centuries after Akbar's death and influenced the architecture of Agra.

Akbar's tomb lies in Sikandra, on the outskirts of Agra. Constructed from the local deep red sandstone and decorated with beautiful calligraphy and geometric patterns, the tomb combines Muslim and native Indian art styles. The tomb is noted for its four white marble minarets topped by *chhatri* (dome-shaped pavilions), which may have inspired similar features in the Taj Mahal. The body of Akbar's favorite wife, Mariam, rests in another elaborate tomb, also in Sikandra. (Polygamy was common in the Mughal Empire.)

But Agra's most prominent tomb, besides the Taj Mahal, is the tomb of I'timād-ud-Daulah. Nicknamed the Baby Taj, it is a direct forerunner of the Taj Mahal. I'timād-ud-Daulah was a Persian-born Mughal official who served as prime minister under Akbar and Mariam's son, Emperor Jahangir (1569–1627), and whose daughter married Jahangir. Built between 1622 and 1628, the tomb signifies an evolution from the first phase of monumental Mughal architecture—primarily built from red sandstone, as in Humayun's and Akbar's tombs—to a new phase, with perhaps an even more pronounced mixing of different architectural traditions. The choice of white marble may have been influenced by Hindu practices "set out in the Vishnudharmottara Purana [a sacred Sanskrit text], which recommended white stone for buildings for the Brahmins," according to the Taj Mahal's official website. (The Brahmins are the priestly class in Hindu society, and the highest-ranking of the four social classes in the traditional caste system.)

Intercultural synthesis was a key characteristic of Agra's Mughal architecture, which mixes Indian, Persian, and Turkish styles, among others. Islam more broadly has a tradition of syncretizing different architectural styles, such as in the Mosque-Cathedral of Córdoba and the Royal Alcázar in Seville. Muslim architects likely drew inspiration from their faith—a famous hadith says, "God is beautiful and loves beauty"—however, they were also constrained by it, since a prominent interpretation of Islam prohibits depicting people or animals. As a result, Muslim artists often avoided sculpting or painting people and animals (with notable exceptions such as the "Persian miniature" painting tradition), instead developing expertise in calligraphy, poetry, and art using abstract geometric patterns. These aniconic designs are among the most distinguishing features of Islamic art and decorate objects of all types, from carpets to stoneware. Alongside calligraphic inscriptions, they also prominently adorn the surfaces of monumental Islamic architecture. Even these distinctive patterns, though, are ultimately the result of cultural intermixing. According to the Metropolitan Museum of Art:

> While geometric ornamentation may have reached a pinnacle in the Islamic world, the sources for both the shapes and the intricate patterns already existed in late antiquity among the Greeks, Romans, and Sasanians in Iran. Islamic artists appropriated key elements from the classical tradition, then complicated and elaborated upon them in order to invent a new form of decoration that stressed the importance of unity and order.

Continuing the virtuous cycle of artistic intercultural borrowing, many notable Islamic geometric designs—like arabesques, or interlaced tendril patterns, and *girih*, or angular knot-like patterns—inspired Christian artists in Italy and elsewhere.

"Arabesque" is, in fact, a French word derived from the Italian term *arabesco*, meaning "in the Arab style." Artistic inspiration flowed in both directions, with Muslim and Christian artists and architects continuously borrowing ideas from one another. For example, the elegant *pietra dura* or *parchin kari* jewel-inlaying technique, mainly developed in Renaissance-era Florence with the generous patronage of the Medici family, was used prominently in Mughal artworks. Agra's Baby Taj made ample use of that inlay technique, but the most elegant use of *pietra dura* in architectural history is widely considered to be in the Taj Mahal itself.

The Taj Mahal was commissioned by Jahangir's son, the grieving Emperor Shah Jahan (1592–1666), for his favorite wife, Mumtaz Mahal (1593–1631), meaning "Jewel of the Palace." *Taj Mahal* is derived from her name. Although Shah Jahan had two other wives, they were consequences of political marriages, and he largely ignored them. The emperor was famously inseparable from Mumtaz Mahal, who accompanied him on his imperial travels and even his military campaigns.

Tragically, even an emperor's family was not safe from the horrifically high rates of child mortality and maternal mortality at the time. Mumtaz Mahal died at age 38 from birth-related complications. Only half of her 14 children survived to adulthood: 4 died in infancy, 1 died at age one and a half, 1 died of smallpox at age three, and another died of smallpox at age seven.

According to legend, as Mumtaz Mahal lay dying, she bound her husband with a promise to build her the most beautiful mausoleum known to man.

The Taj Mahal was built in 22 years by over 20,000 artisans, some summoned from as far as Italy and Persia. The prominent calligraphic adornments are thought to be the work of Amanat Ali Khan Shirazi, the Persian brother of Shah Jahan's prime minister. Ran Mahal, from Kashmir, the northernmost region of the Indian subcontinent, is believed to have designed the complex's magnificent gardens. One disputed theory claims that a Venetian jeweler living in Agra, Geronimo Veroneo, played a part in the design of the Taj Mahal. The main architect was likely Ustad Ahmad Lahauri, a Persian who may have hailed from modern-day Pakistan or Afghanistan. Ustad Isa from Shiraz in the Safavid Empire (modern-day Iran), who may have also been part Turkish, is credited with the site plan. Shah Jahan himself played an active role in the Taj Mahal's design, making "appropriate alterations to whatever the skillful architects had designed after considerable thought and would ask the architects competent questions," according to art historian Ebba Koch.

The Taj Mahal's building materials also came from near and far, with its famous white marble brought from the neighboring state of Rajasthan, the jasper from Punjab, and the jade and crystal adornments from China. Lapis lazuli, carnelian, mother-of-pearl, agate, and emerald were also among the precious gems and stones used in the Taj Mahal's design. The building is thought to have cost about 1 billion 2020 U.S. dollars. Peter Mundy, an Englishman living in Agra at the time, described the construction this way (with spelling modernized for readability):

> This King is now building a sepulcher for his late deceased Queen Taj [Mumtaz] Mahal. . . . He intends it shall excel all other[s]. . . . The building is begun and goes on with excessive labor and cost, prosecuted with extraordinary diligence, gold and silver . . . and marble.

There is widespread international agreement that the Taj Mahal represents the pinnacle of architectural beauty. Type "most beautiful building" into an internet search engine, and chances are the Taj Mahal will appear. The *Encyclopedia Britannica* says, "One of the most beautiful structural compositions in the world, the Taj Mahal is also one of the world's most iconic monuments." *National Geographic* similarly notes, "The Taj Mahal is widely considered one of the most beautiful buildings ever created." The Metropolitan Museum of Art concurs, counting the Taj Mahal "among the most beautiful buildings in the world."

The Taj Mahal's famed tomb forms the centerpiece of a 42-acre complex, which also includes a mosque and a guest house. These architectural wonders stand in a sprawling garden enclosed on three sides by ornate domed and crenelated red sandstone walls. The tomb's main dome is nearly 115 feet high. The palace-like structure is famed for its proportionality, sumptuous attention to detail, and symmetry. It looks the same from all sides, except the one facing the Yamuna River, which was the mourning king's entrance—he would take a barge across the river to pay his respects to his late wife. The acoustics of the Taj Mahal's interior are notable, having, according to the monument's official government website, "a reverberation time (the time taken from when a noise is made until all of its echoes have died away) of 28 seconds providing an atmosphere where the words of the Hafiz, as they prayed for the soul of Mumtaz, would linger in the air." (A *hafiz* is someone who has memorized the Quran.)

Shah Jahan claimed the Taj Mahal's beauty made "the sun and the moon shed tears." He is said to have attempted to make the tomb an earthly replica of the palace he believed Mumtaz would inhabit in paradise. Indian Nobel laureate Rabindranath Tagore described the Taj Mahal as a "teardrop on the cheek of eternity." The Persian poet Kalim Kashani wrote: "It is a [piece of] heaven of the color of dawn's bright face, because from top to bottom and inside out it is of marble. . . . The eye can mistake it for a cloud." The Taj Mahal has also been called "a poem in stone." It is also one of the New Seven Wonders of the World.

The last of the Mughal rulers to commission notable architecture was Shah Jahan and Mumtaz Mahal's son Aurangzeb (1618–1707), who was not especially interested in architecture but had two impressive mosques built, as well as the Bibi Ka Maqbara (Tomb of the Lady) for his wife—which closely resembles the Taj Mahal. Rather than build a separate tomb for his father, Aurangzeb had Shah Jahan interred next to Mumtaz Mahal. (Mumtaz Mahal lies in the center of the Taj Mahal, and Shah Jahan's asymmetrical placement to her side suggests the tomb was originally meant to hold Mumtaz Mahal alone.) Agra's architectural wonders continue to attract thousands of visitors from around the world each year.

While tastes differ, and some may favor different architectural styles—perhaps preferring the Gothic arches of Westminster Abbey in London or the Art Nouveau masterpieces of Antoni Gaudí in Barcelona—there is little doubt that Agra is home to some of the most impressive and visually pleasing architecture ever constructed. Much like the Renaissance paintings of Florence or the classical symphonies of Vienna, the Indo-Mughal architecture of Agra represents a high point of human achievement in the arts. Agra demonstrates the artistic potential of intercultural borrowing and exchange. It is for these reasons that 17th-century Agra towers as a center of progress.

25

Cambridge

PHYSICS

Our next center of progress is Cambridge during the Scientific Revolution. The 16th and 17th centuries constituted a period of drastic change in the way humanity conceptualized and sought to understand the world. Scholars made massive leaps in fields such as mathematics, astronomy, chemistry, and, perhaps most notably, physics. Arguably, no city contributed more profoundly to that new understanding than Cambridge.

Today, Cambridge is a picturesque and walkable university city filled with stunning architecture, cozy pubs, and brilliant minds. "Cambridge is heaven. . . . As you walk round, most people look incredibly bright, as if they are probably off to win a Nobel Prize," says author Sophie Hannah. Indeed, if Cambridge were a country, it would rank fourth on a list of nations with Nobel Prize winners. Cambridge is nicknamed the City of Perspiring Dreams as a nod to its scholars' tireless dedication, in contrast to rival university city Oxford's older nickname, the City of Dreaming Spires.

The great minds that have defined Cambridge are reflected in its architecture and artworks. Architectural highlights include Gothic-style King's College Chapel, featuring the world's largest fan vault, and the Mathematical Bridge, designed in 1749 with the technique of tangential radial trussing, which creates the illusion of an arch although the bridge is built only of straight timbers. The conduit gutters, or runnels, lining many of the city's ancient streets and university buildings owe their construction to none other than Thomas Hobson, the successful stable owner-turned-benefactor from whom we get the term "Hobson's choice" (e.g., "take it or leave it"). The city also boasts fascinating artworks, such as the Corpus Chronophage Clock, a massive "inside-out" electromechanical timepiece that lets onlookers view the usually hidden grasshopper escapement mechanism. The clock is topped with a moving statue called the *chronophage* (time eater), a grasshopper-like entity constructed from stainless steel, gold, and enamel. According to the artist, the horologist and Cambridge alumnus John Taylor (b. 1936), "The time is exactly correct every fifth minute to one hundredth of a second."

Cambridge is also known for the Fitzwilliam Museum, the city's lively market square, and the pastime popular among tourists and students alike of punting (a method of boating in shallow water) on the river Cam—the natural feature around which the city was built. The waterway has made the area an attractive farmsteading site since the Iron Age and has allowed Cambridge to serve as a trading center throughout the ages, including for the Romans (who called the city Duroliponte, meaning "the fort at the bridge"), the Vikings, and the Saxons.

But the true significance of Cambridge began with the founding of the University of Cambridge, which started with a murder mystery. In 1209, a woman was found killed in Oxford, and her death caused an uproar that would alter the course of academic history. She was a local and her fellow townspeople blamed outsiders drawn to their city to study and teach at the University of Oxford. At the time, most of the university's students were teenage clerics, also called clerks. The primary suspect was a liberal arts clerk, who promptly fled his rented home.

The townspeople resented the clerks' special legal privileges, relative wealth, and reputation for drinking and fighting—tense town-versus-gown relations are nothing new. The killing took on an urgent significance in the broader conflict between the townspeople and the university, and a mob of riotous locals soon imprisoned the suspect's roommates. That occurred in the midst of a power struggle between the Church and the crown. The excommunicated King John is said to have personally ordered the hanging of the imprisoned clerks "in contempt of the rights of the church," according to Simon Bailey, then keeper of the archives at Oxford's Bodleian Library, writing for the BBC. The other pupils and instructors at the university fled, fearing further executions. To this day, the crime has not been solved—some say the killing was an accident, whereas others claim it was murder.

The dispersed scholars, possibly including the alleged killer, continued their studies elsewhere. What would become the University of Cambridge started as "no more than a bunch of scholars who had fled Oxford and who had started to teach their students in rented houses in the neighborhood around St. Mary's Church," according to one report based on BBC research. St. Mary's Church today marks the center of Cambridge. In 1214, when the king and the Church had reconciled, Oxford's townsfolk were made to welcome back scholars and offer them reduced rents. But tensions remained high in Oxford (boiling over periodically, such as in the St. Scholastica's Day riot in 1355), and many ex-Oxonians chose to remain in Cambridge.

Soon the University of Cambridge became an intellectual powerhouse in its own right, where great thinkers took human understanding of the world to new heights. "I find Cambridge an asylum, in every sense of the word," the English poet A. E. Housman once quipped. And indeed the city was the birthplace of ideas so groundbreaking that many of them may have sounded mad when they were first expressed.

Cambridge has nurtured great minds in many areas of achievement. Consider the arts. Cambridge's streets have been trod throughout the centuries by literary and poetic geniuses, including Edmund Spenser (1552–1599), Christopher Marlowe (1564–1593), John Milton (1608–1674), William Wordsworth (1770–1850), Lord Byron (1788–1824), Alfred, Lord Tennyson (1809–1892), William Makepeace Thackeray (1811–1863), A. A. Milne (1882–1956), C. S. Lewis (1898–1963), Vladimir Nabokov (1899–1977), Sylvia Plath (1932–1963), and Douglas Adams (1952–2001). Famous Cambridge alumni include comedians, such as John Cleese (b. 1939), Eric Idle (b. 1943), Sacha Baron Cohen (b. 1971), and John Oliver (b. 1977), as well as award-winning actors, such as Emma Thompson (b. 1959) and

Hugh Laurie (b. 1959). And Cambridge gave the world musical feats ranging from the comedic "Always Look on the Bright Side of Life" to the 1980s hit "Walking on Sunshine."

Next, consider philosophy and economics. Cambridge educated the celebrated Catholic theologian, humanist philosopher, and pioneer of religious tolerance Erasmus (1466–1536). Other noted philosophers who were Cantabrigians include Bertrand Russell (1872–1970) and Ludwig Wittgenstein (1889–1951). Cambridge was also the alma mater of influential economists, such as the overpopulation alarmist Thomas Malthus (1766–1834), the father of the beleaguered field of macroeconomics John Maynard Keynes (1883–1946), and Nobel Prize winner Angus Deaton (b. 1945).

But Cambridge's greatest contributions to human progress arguably came in the natural and physical sciences. William Harvey (1578–1657), the physician and anatomist who first detailed the human blood circulatory system, studied at Cambridge. Francis Bacon (1561–1626), the father of empiricism and one of the founders of the scientific method, studied at Cambridge and represented the University of Cambridge (which for a time was a Parliament constituency with its own representatives) in the British Parliament in 1614.

Most historians consider the Scientific Revolution to have begun with the insight of the Polish astronomer Nicolaus Copernicus (who studied in Bologna, another center of progress, and in Padua) that the Earth orbits the sun rather than the sun orbiting the Earth. However, the revolution culminated in the quiet university city of Cambridge with the writing of *Philosophiæ Naturalis Principia Mathematica* (the *Principia*, published in 1687), the groundbreaking work of Isaac Newton (1642–1727) that advanced mankind's knowledge of physics and cosmology. To this day, the University of Cambridge's library retains the first edition of the book owned by Newton, which contains his handwritten notes for the second edition scrawled across it.

If Newton was the father of modern physics, then Cambridge was arguably the field's birthplace. The course of Newton's life revolved around Cambridge; one might say the city's intellectual gravity kept him in its orbit and he could not resist its pull. He received both his bachelor's and master's degrees from the University of Cambridge. Like Bacon, Newton briefly served as a member of Parliament for the University of Cambridge (1689–1690 and 1701–1702). In 1669, only a year after completing his master's degree, Newton became the Lucasian Chair of Mathematics, which is now among the most prestigious professorships in the world, and remained in that position until 1702.

The professorship was made possible thanks to private funding from a benefactor named Henry Lucas (c. 1610–1663). He was a cleric, politician, and Cambridge alumnus who also generously bequeathed a collection of some 4,000 books to Cambridge University Library. Other famous Lucasian Professors include the mathematician Charles Babbage (1791–1871), often called the "father of computing" for conceiving the first automatic digital computer; and the theoretical physicist Stephen Hawking (1942–2018), who—despite severe health challenges from amyotrophic lateral sclerosis (ALS, a progressive motor-neuron illness)—made several notable contributions to his field, including conceptualizing Hawking radiation. The Lucasian Chair has even attracted the attention of popular culture: in the well-known science-fiction franchise *Star Trek*, a central character named Data is said to hold the Lucasian Chair in the late 24th century.

The privately funded professorship allowed Newton to make several breakthroughs in the fields of mathematics, optics, and physics, such as developing the first reflecting telescope. Generous private funding also made the publication of the *Principia* possible. The astronomer and physicist Edmond Halley (1656–1742), the namesake of Halley's comet and the heir to a soap fortune, traveled to Cambridge to encourage, edit, and fund the publication of Newton's *Principia*. In his book, Newton demonstrated how the planets orbit the sun, controlled by gravity. A popular legend holds that Newton first formulated the theory of gravity in the mid-1660s after watching an apple fall from a tree. The precise apple tree often said to have inspired him is, remarkably, still alive—it stands about 70 miles northwest of Cambridge at Newton's family home, Woolsthorpe Manor. Grafted from that storied tree, another "Newton's apple tree" can now be viewed in Cambridge. Whether Newton ultimately did his most momentous thinking in his birthplace or in his intellectual home at Cambridge, one thing is certain: the *Principia* took the world by storm. Its publication is often said to have laid the foundation for modern physics.

After the Scientific Revolution, Cambridge continued to produce world-changing thinkers, such as Henry Cavendish (1731–1810), the discoverer of hydrogen (which he termed "inflammable air"). Later, Cambridge's Cavendish Laboratory was home to major discoveries, including that of the electron in 1897, the neutron in 1932, and the structure of DNA in 1953. The latter came about thanks to the work of physicist Francis Crick (1916–2004) and biologist James Watson (b. 1928), who may have built on findings by other Cantabrigians, including chemist Rosalind Franklin (1920–1958). The physicist Niels Bohr (1885–1962), who developed the Bohr model of the atom, also studied at Cambridge. The university city was the site of other groundbreaking moments in scientific history, such as the invention of in vitro fertilization technology (1968–1978), the first identification of stem cells (1981), and the earliest eye-recognition technology (1991).

It also must be mentioned that Cambridge educated the founder of evolutionary biology, Charles Darwin (1809–1882). He forever altered the understanding of living things by positing the fundamental scientific concepts of animal and human evolution and natural selection. Along with Newton, he is probably the most influential figure in scientific history to emerge from Cambridge's classrooms—and one of the most influential men in history, full stop.

Cambridge grew from its unconventional murder-mystery origins into an intellectual center that played a pivotal role in the Scientific Revolution, which is often said to have been made complete with the publication of the *Principia*. Thanks to the rigorous culture of Cambridge's academic community and funding from generous benefactors, the city has often served as the headquarters of mankind's quest for truth and understanding. Many scholars believe that the new way of thinking that emerged during the Scientific Revolution directly led to the Enlightenment movement in the 17th and 18th centuries. The innovations of the Scientific Revolution continue to form the basis for humanity's present understanding of the natural world, including modern physics. It is for these reasons that we must gravitate toward Cambridge as a center of progress.

26

Paris

ENLIGHTENMENT

Our next center of progress is Paris, sometimes called the "center stage" or "home" of the Enlightenment. The salons and coffeehouses of 18th-century Paris provided a place for intellectual discourse where the *philosophes* birthed the so-called Age of Enlightenment. The Enlightenment was a movement that promoted the values of reason, evidence-based knowledge, free inquiry, individual liberty, humanism, limited government, and the separation of church and state. Although a long-distance intellectual community, known as the Republic of Letters (*Respublica literaria*), fostered communication among intellectuals across borders and oceans, Paris nonetheless served as an important geographical center of intellectual life. As Paris became known for its intellectuals' challenging of traditional beliefs, it earned the nickname the City of Light (*la Ville Lumière*). It is undeniable that the city's thinkers and the broader Enlightenment movement altered history. Some scholars such as Harvard University psychologist Steven Pinker credit Enlightenment values with much of the scientific and moral progress humanity has made since then.

Today, Paris remains France's capital and most populous city, with more than 2 million residents. The city continues to serve as a significant center of diplomacy, commerce, high fashion, cuisine, science, and the arts, as it has done since at least the 17th century. Paris is one of the world's top tourist destinations, famed for its architectural landmarks, museums, restaurants, and charming atmosphere. For its romantic reputation, one of the city's nicknames is the City of Love. Paris is thus also a popular destination for weddings and honeymoons. The mystique that Paris holds in the public imagination is difficult to encapsulate, but the words of the Nobel Prize–winning theoretical physicist Walter Kohn perhaps best sum up the aspect of the city that concerns us: "Paris somehow lends itself to conceptual new ideas. . . . There is a certain magic to that city."

The site where Paris now stands has been inhabited since around 7600 BCE. The city's museums maintain archaeological artifacts dating to the Stone Age and the Roman Empire, although those are not periods of history that are commonly associated with Paris. What began as a small settlement on the river Seine's banks, grew rapidly in population and political importance. Paris gets its name from an Iron Age Celtic tribe, the Parisii, who fortified the area circa 225 BCE. In 52 BCE, the Romans conquered the site and named it Lutetia Parisiorum (Marsh of the Parisii). By the third century CE, local Germanic tribes challenged the Roman rule of the city. By the end of the fifth century, Paris came under the complete control of the Franks, a confederation of Germanic tribes. In 508, the Franks made Paris their capital. In 843, the Kingdom of Francia split, with East Francia becoming the predecessor state to Germany and West Francia becoming the earliest iteration of the Kingdom of France. As France's political influence expanded over the centuries, Paris became an important economic and cultural hub.

In the 18th century, the center of cutting-edge intellectual discourse shifted, in part, from universities to coffeehouses and salons, where controversial thinkers could find financial backing. Europe's adoption of coffee substituted the continent's constant consumption of alcohol, a depressant, with caffeine, a stimulant, and coffeehouses became hubs for the debate of politics and philosophy. Women played an important, if unequal, role in the Enlightenment. Wealthy or well-connected women known as *salonnières*—such as Marie Thérèse Rodet Geoffrin (1699–1777), who is sometimes called the inventor of the Enlightenment salon—hosted the era's upscale intellectual gatherings. Even upper-class women at the time were typically denied formal educational opportunities, but the salons served as a socially acceptable way for women to engage in intellectual life. Other key salonnières included Jeanne Julie Éléonore de Lespinasse (1732–1776) and Suzanne Necker (1739–1794), the Swiss wife of King Louis XVI's finance minister.

In Enlightenment-era salons, nobles and other wealthy financiers intermingled with artists, writers, and philosophers seeking patronage and opportunities to discuss and disseminate their work. The gatherings gave controversial philosophers, who would have been denied the intellectual freedom to explore their ideas in the academy, the freedom to develop their critiques of existing norms and institutions. Influential Parisian and Paris-based thinkers of the period included the Baron de Montesquieu (1689–1755); François-Marie Arouet, better known by his pen name Voltaire (1694–1778); the Genevan expat Jean-Jacques Rousseau (1712–1778); and the writer Denis Diderot (1713–1784).

The salons were famous for sophisticated conversations and intense debates; however, it was letter writing that gave the philosophes' ideas a wide reach. A community of intellectuals that spanned much of the Western world—known as the Republic of Letters—increasingly engaged in the exchanges of ideas that began in Parisian salons. Thus, the Enlightenment movement based in Paris helped spur similar radical experiments in thought elsewhere (such as the Scottish Enlightenment that is the subject of the next chapter).

Thanks in part to funding and feedback provided by salon patrons, Paris's philosophes put many of their ideas to paper. In 1748, Montesquieu published *Spirit of the Laws*, which advocated a separation of governmental powers. He posited that no single branch or part of government should have too much power relative to any other, which at the time was a groundbreaking proposition.

In 1751, Diderot helped create the *Encyclopédie*, among the first modern, general-purpose encyclopedias. Over 27 years, he served as its chief editor and helped produce 28 different volumes of the *Encyclopédie*. The controversial book series was banned by both the Catholic Church and the French government, so he had to go into hiding to complete the final volumes. The production of the *Encyclopédie* has been called the seminal intellectual achievement of the French Enlightenment.

In 1759, Voltaire published his best-known work, *Candide*, a sarcastic novella that was also widely banned for its criticisms of religious and political institutions. Although Parisian by birth, Voltaire spent relatively little time in Paris because of frequent exiles occasioned by the ire of French authorities. Voltaire's time hiding out in London, for example, enabled him to translate the works of the political philosopher and "father of liberalism" John Locke, as well as the English mathematician and physicist Isaac Newton.

In 1762, Rousseau published *The Social Contract*, which argued, among other things, that laws should reflect the "will of the people" and that monarchs have no "divine right" to rule. This work, too, was censored. His ideas proved influential among the leaders of the French Revolution a generation later (1789–1799). That said, some scholars consider Rousseau to have been a counter-Enlightenment figure because of his skepticism of modern commercial society and highly romanticized view of primitive existence.

The Enlightenment blossomed in 18th-century Paris largely thanks to the efforts and generosity of private individuals. The authorities made many attempts to stifle new ideas that challenged the existing order. The French government often censored or banned writings and exiled intellectuals. But private funding, funneled to trailblazing thinkers through the salons, allowed new ideas to take root and thrive.

Other cultural developments were also occurring in Paris at the time. The city is the birthplace of haute cuisine and restaurants. During the 1760s and 1770s, the first modern restaurants emerged in France. In 1782, the pastry chef to the future Louis XVIII, Antoine Beauvilliers (1754–1817), opened the first prominent fine-dining establishment in Paris. As the monarchy weakened and more court chefs left their posts to open restaurants, the nouveaux riches helped sustain the new establishments and fund the development of the culinary arts. French cuisine remains a significant cultural achievement that continues to be a source of pride for Parisians.

During the 18th century, Paris was also a center of music and opera, painting (particularly within the baroque, rocaille, and neoclassical artistic movements), and fashion (as the city had been for a century thanks to the elaborate clothes of King Louis XIV's court). But it was the new ideas of the Enlightenment that defined the city in that era and ultimately transformed the world. Enlightenment ideals helped bring about the French Revolution, which caused horrific bloodshed and chaos, but also showed that it was possible to rethink centuries-old institutions.

In other words, by providing a home base for the Enlightenment and the far-ranging Republic of Letters, Paris helped spread new ideas that would ultimately give rise to new forms of government. The Enlightenment ideals of republicanism, separation of government powers, separation of church and state, and respect for civil liberties helped animate the French and American revolutions. The Enlightenment emphasis on reason and evidence helped lay the groundwork for life-changing innovations in science and technology. The Enlightenment, in some ways, helped pave the way for the later Industrial Revolution, a turning point in human history that created unprecedented wealth and eventually raised living standards to previously unimaginable highs.

Thanks to new technologies, the city's nickname, the City of Light, developed a double-meaning, as Paris became among the first cities to install gas street lighting along its boulevards and monuments in the 19th century. Between 1853 and 1870, Paris installed some 15,000 gas streetlights. The 19th century also saw Parisian artistic achievements reach new highs with the erection of architectural marvels such as the iconic Eiffel Tower and the production of impressionist and postimpressionist masterpieces. There were too many groundbreaking France-based painters of the era to name. Still, some of the more influential ones include Claude Monet (1840–1926), Paul Cézanne (1839–1906), Edgar Degas (1834–1917), Édouard Manet (1832–1883), Pierre-Auguste Renoir (1841–1919), Georges-Pierre Seurat (1859–1891), Henri Rousseau (1844–1910), and Vincent van Gogh (1853–1890). That era also saw new heights of French literature with well-known writers such as Victor Hugo (1802–1885), Honoré de Balzac (1799–1850), and Alexandre Dumas (1802–1870).

Today, Paris continues to be renowned the world over as a center of high culture. However, it no longer has the same reputation it did in the 18th century as the world's intellectual capital.

Through a flourishing of private funding for new ideas, including controversial ones, 18th-century Paris became the cradle of the Enlightenment and the geographical base of the far-ranging Republic of Letters. As noted, the city has made many other notable achievements, particularly in the arenas of high culture, including painting, music, clothing design, and the culinary arts. But the city's greatest contributions to human progress were the world-altering ideas that emerged among Paris's thinkers during the Age of Enlightenment. Paris thus shines as a center of progress.

27
Edinburgh
SOCIAL SCIENCE

Our next center of progress is Edinburgh. The city was at the heart of the Scottish Enlightenment—a vital period in intellectual history that spanned the 18th and early 19th centuries. The thinkers of the Scottish Enlightenment made important breakthroughs in economics, mathematics, architecture, medicine, poetry, chemistry, theater, engineering, portraiture, and geology.

Today, Edinburgh remains Scotland's intellectual and cultural center, as well as its capital. The city's name comes from an old Celtic word, *Eidyn*, which is a name for the area, and *burgh*, which means "fortress." A hilly city on Scotland's east coast, Edinburgh is home to a famous castle dating to at least the 12th century. Edinburgh Castle is Scotland's most visited tourist attraction, drawing over 2 million visitors in 2019 alone. The city is also home to the University of Edinburgh, one of Scotland's most prestigious universities. Edinburgh's nicknames include Auld Reekie (Old Smoky) for Old Town's smoky chimneys. The city is also sometimes called Auld Greekie, or the Athens of the North, for the city's role as a hub of philosophy. Edinburgh's medieval Old Town and neoclassical New Town together compose a single UNESCO World Heritage site.

Archaeological evidence suggests the area where Edinburgh now stands has been inhabited since at least 8500 BCE. Celtic tribes were the main inhabitants. Over the centuries, the area was ruled by various peoples, including Welsh-speaking Brittonic Celts. Edinburgh came under Scottish rule around 960 CE, when King Indulf the Aggressor seized the settlement. Edinburgh became the Scottish capital in 1437, replacing Scone.

Scotland in the 18th century had just undergone decades of political and economic turmoil. Disruption was caused by the House of Orange's ousting of the House of Stuart, the Jacobite rebellions, the failed and costly colonial Darien scheme, famine, and the 1707 union of Scotland and England. Yet Scotland, particularly Edinburgh, was to embark on an exciting new journey.

If you could visit Edinburgh during the Scottish Enlightenment, you would enter a cold, compact, walled-off city of winding, cobblestone streets. The Scottish author James Buchan has described the city of the era as "inconvenient, dirty, old-fashioned, alcoholic, quarrelsome and poor." But through the fog, you would see the warm glow of lights in the windows of the university buildings, the homes hosting reading societies and club meetings, and the taverns serving haggis and whisky to patrons discussing philosophy. The city was alive with the energy of new ideas and the spirit of scientific inquiry. Although Edinburgh was then a city of merely 40,000 residents, it was crowded with great minds tackling big questions.

It helped that the city's religious culture was welcoming to new ideas. The dominant Presbyterian Church had just undertaken a successful literacy campaign. Scotland, then one of western Europe's poorest countries, enjoyed perhaps the world's highest literacy rate. The reigning faction within the Presbyterian Church comprised moderate, open-minded clergymen. Those moderates formed close ties to many of the Scottish Enlightenment's key figures and encouraged their work. There was also a more conservative faction within the Presbyterian Church that disdained the Enlightenment scholars' work and even tried to excommunicate the philosopher David Hume (1711–1776) for heresy. The better-connected moderate faction within the Church shielded Hume from excommunication.

The moderate Presbyterian reverend William Robertson (1721–1793) became the University of Edinburgh's principal and founded one of the Scottish Enlightenment's most prominent intellectual societies in 1750. (The principal, then as now, was responsible for the overall operation of the university.) Robertson's Select Society of Edinburgh counted among its members such luminaries as Hume, the philosopher and historian Adam Ferguson (1723–1816), and the economist Adam Smith (1723–1790). In 1783, Hume's fiercest philosophical opponent, Thomas Reid (1710–1796), cofounded the Royal Society of Edinburgh, another intellectual society.

Much like the French Enlightenment's Parisian salons, the numerous reading societies and intellectual men's clubs that sprang up throughout Edinburgh enabled the city's success. Unlike in Paris, where women often hosted salons, sexist cultural norms excluded women from Edinburgh's intellectual gatherings, with rare exceptions, such as the poet and socialite Alison Cockburn (1712–1794). A modern woman would not wish to live in 18th-century Edinburgh, but the men at the time found the networking and debate opportunities afforded by the city's various clubs invaluable. The French writer Voltaire opined in 1762, "Today it is from Scotland that we [Europeans] get rules of taste in all the arts, from epic poetry to gardening."

Scotland made its mark in the literary realm, producing such figures as the inimitable poet Robert Burns (1759–1796) and the Edinburger novelist Sir Walter Scott (1771–1832). Scotland also pioneered new landscaping, architectural, and interior design tastes. That was thanks largely to the Edinburgh-raised and -educated architect Robert Adam (1728–1792). Together with his brother James (1730–1794), he developed a new approach to architecture known as the "Adam style." The Adam style influenced many residences in 18th-century England, Scotland, Russia, and the United States after Independence, where it evolved into the so-called Federal style. Scotland also led the way in portrait painting, thanks to taste-making Edinburger artists such as Allan Ramsay (1713–1784) and Sir Henry Raeburn (1756–1823).

Although the Scottish Enlightenment produced many contributions to the arts and humanities, it also gave rise to groundbreaking work in the sciences. Thomas Jefferson, in 1789, wrote, "So far as science is concerned, no place in the world can pretend to competition with Edinburgh." The Edinburger geologist James Hutton (1726–1797) redefined his field by developing many of the fundamental principles of his discipline. The chemist and physicist Joseph Black (1728–1799), who studied at the University of Edinburgh, discovered carbon dioxide, magnesium, and the important thermodynamic concepts of latent heat and specific heat.

The physician William Cullen (1710–1790) helped make Edinburgh Medical School into the English-speaking world's leading medical education center. There, he helped train many notable scientists, including Black and the anatomist Alexander Monro Secundus (1733–1817). The latter was the first person to detail the human lymphatic system. Sir James Young Simpson (1811–1870), admitted to the University of Edinburgh at the young age of 14, went on to develop chloroform anesthesia. That invention vastly improved the experience of surgical patients. It also saved Queen Victoria (1819–1901) and countless other women from unnecessary suffering during childbirth.

The Scottish Enlightenment also advanced mathematics and engineering. The mathematician and University of Edinburgh professor Colin Maclaurin (1698–1746), a child prodigy who entered university at age 11, made notable contributions to the fields of geometry and algebra. The civil engineer Thomas Telford (1757–1834), who worked for a time in Edinburgh, was so prolific that he earned the nickname the Colossus of Roads (a play on one of the Seven Wonders of the Ancient World, the Colossus of Rhodes). The Scottish engineer and inventor James Watt (1736–1819) greatly improved the steam engine's design and thus helped bring about the Industrial Revolution.

American author Eric Weiner has argued that the key to Edinburgh's sudden, unexpected success was Scottish practicality. The *Encyclopedia Britannica*—which was founded in Edinburgh in 1768 and was thus an invention of the Scottish Enlightenment—also claims that underlying the city's diverse achievements were several notable developments in Scottish philosophy, all of which had a practical bent. Those developments were skepticism toward the so-called rationalist school of thought (which held that all truths could be deduced through the use of reason alone), a focus on empirical methods of scientific inquiry, the emergence of a philosophy of "common sense," and attempts to develop a science of human nature.

Popularization of empiricism was among the greatest contributions of the Scottish Enlightenment to human progress. Relatedly, "common-sense realism," advanced by thinkers such as Ferguson, emphasized real-world observations rather than abstract theorizing and held that the uneducated, common man was an intellectual's equal in matters of basic common sense. Common-sense realism influenced the thinking of the U.S. Founding Fathers Thomas Jefferson and John Adams, among others. Hume's *A Treatise of Human Nature* (1739), among the most influential philosophical works in history, was the foundational text of cognitive science.

The desire to understand human behavior gave rise not only to cognitive science, but also to economics. Adam Smith is widely regarded as the founder of modern economics. His

An Inquiry into the Nature and Causes of the Wealth of Nations (1776) was among the first works to delve into such topics as the division of labor and the benefits of free-trade economies (as opposed to mercantilism and protectionism). The work not only influenced economic policy soon after it was published, but also helped define the terms of economic debate for centuries. Every important economic thinker since Smith, including those who strongly disagreed with him, such as Karl Marx, nonetheless cited the Scotsman and wrestled with his ideas.

By creating the field of economics, Smith helped humanity think about policies that enhance prosperity. Those policies, including economic freedom, for which Smith advocated, have since helped raise living standards to heights that would have been unimaginable to Smith and his contemporaries. (Explore the evidence for yourself on websites such as HumanProgress.org.)

Edinburgh was an improbable center of progress. A relatively small, unkempt, and inhospitable locale emerged from a century of instability to take the world by storm. Widespread literacy, open-mindedness, intense debates at intellectual gatherings, and a practical grounding aided the city's successes. Edinburgh was essentially a small university town that punched far above its weight in human achievement. The American Founding Father Benjamin Franklin noted, "The University of Edinburgh possessed a set of truly great men . . . as have ever appeared in any age or country." For its innumerable achievements, and particularly for giving humanity empiricism and economics, Scottish Enlightenment–era Edinburgh is rightfully a center of progress.

28

Philadelphia

LIBERAL DEMOCRACY

Our next center of progress is Philadelphia, nicknamed the Cradle of Liberty and the Birthplace of America. This early U.S. capital is where the Second Continental Congress signed the Declaration of Independence. It is also where a new form of government was debated and put into practice. Previously, the prevalent form of political organization was monarchy. But the Founders of the American republic attempted to create something new.

Today, Philadelphia is the largest city in Pennsylvania and forms the heart of the seventh-largest metropolitan area in the country. The city is a major cultural center, known for its historical monuments such as the Liberty Bell, its famous cheesesteak sandwiches, the University of Pennsylvania, and cultural icons such as the famous "Rocky Steps." The historic Independence Hall—where the Declaration of Independence and the Constitution (which succeeded the Articles of Confederation) were signed—is a UNESCO World Heritage site. "The principles debated, adopted and signed in Independence Hall have profoundly influenced lawmakers and policymakers around the world," according to UNESCO.

William Penn (1644–1718), an English Quaker, founded Philadelphia in 1682 as the capital of his new "Pennsylvania Colony." The city's name means "brotherly love" in Greek. It pays homage to an ancient city, in what is today Turkey that is referenced in the Bible. Ancient Philadelphia served as an early center of Christianity. The Quakers, a Protestant denomination, were known for promoting pacifism and for their opposition to slavery. The latter was a particularly radical position at the time. Initially, about 7 percent of Philadelphia households owned slaves. It is estimated that by 1767 that figure had grown to 15 percent of Philadelphia's households. In 1712, the Pennsylvania Assembly—which met in Philadelphia—banned the import of slaves into the colony. That decision was overruled by the British government under Queen Anne in early 1713. The next year, 1714, and again in 1717, the Pennsylvania Assembly tried to limit slavery in the colony. Each time, the British government in London rejected the decision.

Penn founded the Pennsylvania colony as a "Holy Experiment" to be governed by Quaker values. Its laws differed from those in the other American colonies in notable ways. Pennsylvania guaranteed religious freedom, promoted education for girls as well as boys, and sought to rehabilitate prisoners by teaching them a trade, rather than simply punishing the offenders. The death penalty in Pennsylvania was reserved for those convicted of murder or treason at a time when, in Britain, people were put to death for a wide variety of trivial offenses. Penn—who kept at least 12 slaves in contradiction to his public stance on slavery—proposed before the Pennsylvania Assembly legislation that would have freed Pennsylvania's slaves and given the latter property in a new township. Alas, his proposal was voted down.

Abolitionism, universal education, and enlightened penal practices were not the only radical ideas spreading through Philadelphia in the 18th century. Many colonists grew increasingly frustrated with their lack of political representation in far-off, yet micromanaging, Britain. Enlightenment ideas inspired the discontented colonists to embark on an experiment that would change the world. In 1774, representatives from 12 of the 13 British colonies in America convened in Philadelphia. They formed the First Continental Congress. (The colony of Georgia did not dare send a representative as it was struggling in a war against local tribes and could not risk losing British military assistance.)

The First Continental Congress endorsed the boycotting of British goods and militia raising, but its most significant decision was to call for a Second Continental Congress. Although no war against Britain was yet officially declared, George Washington (1732–1799), one of the delegates from Virginia, bought new muskets and military apparel. He also placed an order for a book on military discipline. As he walked the cobblestone streets of Philadelphia, the future president sensed that war was imminent.

Several events escalated the conflict. In 1775, British forces attempted to seize a Massachusetts armory. Local militiamen resisted. It is unclear which side fired first, but the resulting violence left 90 Americans and 273 Britons dead. Americans then besieged the British-held city of Boston. Those events—the Battles of Lexington and Concord, and the Battle of Bunker Hill—are often considered the start of the American Revolution.

However, at that point, the conflict between the colonists and the British still resembled a civil war, not a revolution. Many colonists wanted a resolution to the violence that did not involve separating from Britain. Rather, they wanted to receive better political representation in the British Parliament. In January 1776, the English-born American writer Thomas Paine (1737–1809) published a pamphlet titled *Common Sense* that argued for independence from Britain and for the formation of a liberal democratic republic. Paine published that work in Philadelphia and it soon sold more than 100,000 copies. It energized public support for a break from Britain and experimentation with the republican form of government. Founding Father and second U.S. president John Adams (1735–1826) famously opined, "Without the pen of the author of *Common Sense*, the sword of Washington would have been raised in vain." The printing presses of Philadelphia thus catalyzed the American Revolution.

Philadelphia then hosted the Second Continental Congress. Although the Second Continental Congress met in several other places as well, it was in Philadelphia that the Congress adopted the Declaration of Independence. A Virginian, Thomas Jefferson (1743–1826), drafted the document while staying at a brick mason's house in Philadelphia. The document laid out the rebel colonists' reasoning for wishing to separate from Britain and spelled out several ideals of the new nation. The United States of America became the first country founded on Enlightenment principles, including human rights and consensual government. The Declaration's most well-known passage reads:

> We hold these Truths to be self-evident, that all Men are created equal, that they are endowed by their Creator with certain unalienable Rights, that among these are Life, Liberty, and the pursuit of Happiness—That to secure these Rights, Governments are instituted among Men, deriving their just Powers from the Consent of the Governed.

Many of the ideas expressed in the document came directly from Enlightenment philosophers. For example, it paraphrased the "father of liberalism" John Locke's belief in the rights to "life, liberty, and property." The young American republic did not always live up to its own ideals—most glaringly in the case of slavery. The Founding ideals have nonetheless inspired countless Americans to strive to create a freer society with greater legal equality. The country's Founding values thus ultimately helped bring about the end of slavery (1865), the expansion of the voting franchise to all races (1870) and women (1920), and the right to marry for interracial couples (1967) and same-sex couples (2015). In other words, the Declaration of Independence's eloquent statement of Enlightenment ideals has continued to resonate across generations and to encourage progress.

It is unsurprising that Philadelphia served as the headquarters, if not always the official capital, of the new nation during the war. It was the young country's most populous city. As with so many other centers of progress, a relatively large population helped the city thrive and act as a cultural hub. Although Philadelphia had only about 40,000 residents, it would have felt crowded compared with other towns in the colonies. Philadelphia was also the country's busiest port. If you could visit Philadelphia during the American Revolution, you would enter a prospering city of shops and brick rowhouses, abuzz with the latest news about the war.

You might have run into the scientist, newspaperman, and statesman Benjamin Franklin (1706–1790), one of the most prominent proponents of the revolution. He also helped shape Philadelphia. He first moved from his hometown of Boston, governed by Puritans, to the relatively tolerant Philadelphia at the age of 17, to seek work in the printing industry. (He had previously apprenticed for his brother's newspaper, which the Boston authorities soon banned.) In 1729, Franklin began the *Pennsylvania Gazette*, which became one of the top papers in the colonies. He founded Philadelphia's Library Company in 1731, thus pioneering the concept of a lending library at a time when books were often prohibitively expensive. Membership subscriptions funded the library. In 1751, Franklin also founded the Pennsylvania Hospital, funded by charity (including financial support from many of Philadelphia's wealthiest families) and a grant that Franklin secured from the government to match private donations. The hospital served patients free of charge, and Philadelphia soon became the medical capital of the colonies that would later become the United States.

Once the revolution began, the threat of seizure by the British loomed large in Philadelphians' minds. In the autumn of 1777, those fears came to pass. Writer Patrick Glennon calls the British occupation of the city, "one of the greatest blunders of the Revolutionary War." The capital moved to Baltimore during the occupation. As the Philadelphians suffered from wartime shortages, the occupying British officers gained a reputation for living in luxury and for looting. Elizabeth Drinker (c. 1735–1807), a Quaker diarist residing in Philadelphia at the time, described the situation: "How insensible do these people appear, while our Land is so greatly desolated, and Death and sore destruction has overtaken and impends over so many." In 1778, as the American forces grew stronger thanks to aid from France, the British recalled their troops from Philadelphia. In 1783, the war ended in a victory for the rebels.

Toward the end of the American Revolution, Pennsylvania abolitionists—including many Quakers and Presbyterians motivated by their religious values—helped abolish slavery in Pennsylvania by passing legislation in Philadelphia in 1780 that phased out the practice. Soon after, several other U.S. states (New Hampshire, Connecticut, and Rhode Island) followed suit with legislation modeled closely after Pennsylvania's. (Vermont had banned slavery in 1777, becoming the first colony to ban it outright.) Continuing its central role in the young republic, Philadelphia served as the official U.S. capital between 1790 and 1800 while Washington, DC, was constructed.

By being the "cradle of liberty" and headquarters of the American Revolution, Philadelphia helped humanity discover the benefits of liberal democracy. The ideas at the heart of the new form of government proved so successful that today, representative liberal democracies can be found throughout much of the world. Philadelphia was also a notable early center of anti-slavery abolitionism, Enlightenment values, medical science, and culture. It is for these reasons that Philadelphia rings true as a center of progress.

29

Vienna

MUSIC

Our next center of progress is Vienna, nicknamed the City of Music. From the late 18th century through much of the 19th century, the city revolutionized music and produced some of the Classical and Romantic eras' greatest works. The sponsorship of the then-powerful Habsburg dynasty and the aristocrats at Vienna's imperial court created a lucrative environment for musicians, attracting the latter to the city. Some of history's greatest composers—including Ludwig van Beethoven, Johannes Brahms, Joseph Haydn, Franz Schubert, and Wolfgang Amadeus Mozart—lived and created music in Vienna. Many of history's most significant symphonies, concertos, and operas thus originated in Vienna. Even today, pieces composed during Vienna's golden age continue to dominate orchestral music performances worldwide.

Today, Vienna is the capital and most populous city in Austria, with nearly 2 million residents. The city is famous for its cultural icons, including multiple historic palaces and museums, as well as its coffeehouses, upscale shops, and high quality of life. The historical city center is a UNESCO World Heritage site. The city influenced the name of the Austrian school of economics, an influential school of economic thought emphasizing individual choice that emerged there in the late 19th and early 20th centuries, led by economists such as Carl Menger (1840–1921), F. A. Hayek (1899–1992), and Ludwig von Mises (1881–1973). Although Vienna's contributions to the field of economics are substantial, the city is best known for music and still bills itself as the World Capital of Music, hosting numerous concerts. In addition to its historical role in revolutionizing music, Vienna has continued to inspire musicians in more recent times. Vienna's official tourism website notes that the city is the subject of more than 3,000 songs, including two by former Beatles and the eponymous Billy Joel hit.

The site next to the Danube River where Vienna now stands has been inhabited since at least 500 BCE when evidence suggests that ancient Celts lived in the area. Circa 15 BCE, the site became home to a Roman fort. Vienna's location along the Danube made it a natural trading hub. Coins from the Byzantine Empire made their way to Vienna by the sixth century CE, indicating that the city engaged in far-reaching trade. By 1155, Vienna became the capital of the Margraviate of Austria, which was upgraded to a dukedom the following year. Throughout the centuries, the area continued to grow in wealth and political importance. In the mid-15th century, Vienna became the Habsburg dynasty's headquarters and the de facto capital of the Holy Roman Empire. The Habsburgs were once among the most influential royal families in Europe. Albeit with greatly diminished power, the family remains active in politics to this day. (As an interesting bit of trivia, the current head of the Habsburg family was the first royal person to contract COVID-19.)

As an increasingly prominent center of trade and culture, the city became a target for military attacks and vulnerable to foreign diseases. Vienna weathered Hungarian occupation in the 15th century, attempted Ottoman invasions in the 16th and 17th centuries, and a devastating epidemic (likely bubonic plague) in 1679 that killed a third of the city's inhabitants. To this day, an ornate sculpture-covered column celebrating the epidemic's end can be viewed in the city center. In 1804, as the Napoleonic Wars raged, Vienna became the new Austrian Empire's capital. Despite its encounters with war and disease, Vienna's status as a site of high culture only grew.

The Habsburg family and the imperial court sought to increase their prestige by funding the arts, particularly music. With strong ties to Italy and the Catholic Church, as early as the 17th century, the Habsburgs brought over 100 Italian musicians to Vienna and introduced cutting-edge Italian musical innovations such as opera and ballet to the city, as well as increasingly extravagant productions of sacred music. As part of the counter-Reformation, the Catholic Church promoted grand musical and artistic projects.

In 1622, the head of the Habsburg family, the Holy Roman emperor Ferdinand II (1578–1637), married the music-loving Eleonora, a princess of Mantua (1598–1655). Empress Eleonora's artistic patronage is credited with making the Viennese court a center of baroque music and fledgling theatrical forms such as opera. As the Habsburgs financed increasingly lavish musical performances to celebrate family occasions like birthdays and grandiose religious music performances, the monetary incentive drew more and more musicians to the city from across Europe. By the 1760s, music was so embedded in Vienna's culture that members of not only the nobility but also the prosperous middle class began to act as music patrons.

The Austrian composer Joseph Haydn (1732–1809), often called the "father of the symphony" and the "father of the string quartet," rose from humble origins as the son of a wheelwright and a cook to become Europe's most celebrated composer for a time. Haydn did his early work as a court musician for a rich family in a remote estate but was eventually drawn to Vienna, where he was lavished with funding and treated as a celebrity. Haydn's magnum opus, *The Creation*, an oratorio celebrating the biblical book of Genesis, first premiered in a private performance for a society of Vienna's music-loving nobles. *The Creation* publicly debuted in Vienna's Burgtheater in 1799 and sold out long before the performance. While in Vienna, Haydn became a mentor to Mozart (1756–1791) and tutored Beethoven (1770–1827).

The son of a Salzburg music instructor, Wolfgang Amadeus Mozart first performed in Vienna's Schönbrunn Palace when he was just 6 years old, alongside his 10-year-old sister. The Habsburg empress Maria Theresa (1717–1780) paid the siblings

100 gold ducats and gave them expensive garments in thanks. Mozart is widely considered to be one of the greatest composers of all time. He enjoyed the most financially successful part of his career while in Vienna. There, he and his wife rented an elegant apartment, bought expensive furniture, had multiple servants, sent their son Karl to a prestigious school (in Prague), and lived a luxurious lifestyle in general. Maria Theresa's son and successor, Joseph II (1741–1790), appointed Mozart to the position of court chamber music composer, giving the latter a salary on top of his earnings from his concerts and other patrons.

However, in his later years, Mozart suffered financially. As the Austro-Turkish War (1788–1791) raged and reduced the prosperity of Vienna and its aristocrats, funding for musicians became harder to secure. Even as his earnings declined, Mozart's expenses remained high and he fell into debt. He had begun to recover financially by finding new patrons outside Vienna when, at the age of 35, he died of a sudden illness that may have been influenza or a streptococcal infection (or, some believe, poison). One of his greatest masterpieces, *Requiem*, remained unfinished. Adding to the piece's mystique, Mozart's widow famously claimed that a mysterious stranger had commissioned *Requiem* and that Mozart felt he was composing the mass for his own funeral.

Beethoven, also among the most beloved composers in history, moved from Bonn to Vienna at age 21. He quickly gained a positive reputation as a pianist and became a favorite at the imperial court. Among his most prominent patrons was Archduke Rudolf (1788–1831), a cardinal in the Catholic Church and a member of the Habsburg family. Beethoven's most profitable concerts were repeat performances of his work celebrating the Duke of Wellington's defeat of Napoleon (op. 91) and his Seventh Symphony (op. 92), which was also inspired by the Napoleonic Wars. Beethoven's achievements were all the more impressive after he became mostly deaf in his later years, but continued to compose innovative music. What is widely heralded as Beethoven's greatest work, his Ninth Symphony (op. 125), premiered in Vienna in 1824. It remains one of the most performed musical pieces the world over.

Schubert (1797–1828), a Vienna native, produced an acclaimed body of work within his short life thanks to the patronage of the city's aristocracy. His greatest work, *Winterreise* (Winter Journey), took its lyrics from a series of poems by Wilhelm Müller and explored themes of loneliness and suffering. He died at age 31, probably of typhoid fever or perhaps syphilis.

Brahms (1833–1897), born in Hamburg, also worked for most of his professional life in Vienna. His Fourth Symphony is often listed among his best works. Brahms believed in "absolute music," or music that is not "about" anything in particular and does not explicitly reference any specific scene or narrative. However, some scholars believe the Fourth Symphony may have been inspired by Shakespeare's play *Antony and Cleopatra*.

After the Classical and Romantic music eras, Vienna continued to serve as a major center of cultural innovation. Vienna was at the center of an art nouveau movement in the 20th century and produced famous artists such as Vienna-born Gustav Klimt (1862–1918). But Vienna remains best known for its musical accomplishments in the 18th and 19th centuries.

Music has enlivened human existence since prehistory. Radiocarbon dating suggests that flutes excavated in Germany, and carved of mammoth ivory, are between 42,000 and 43,000 years old. The oldest surviving written melody, preserved on a clay cuneiform tablet, is an ode to an ancient goddess of orchards, first composed circa the 14th century BCE. The oldest intact and translated surviving musical composition with both lyrics and a melody may date as far back as 200 BCE and is written in ancient Greek. It is engraved on a marble column marking the grave of a woman named Euterpe (literally, "rejoicing well"). She was, appropriately, named after the Muse of music. The lyrics of the song, thought to be written by Euterpe's widower, read in translation:

Ὅσον ζῆς φαίνου	hóson zêis, phaínou	While you live, shine
μηδὲν ὅλως σὺ λυποῦ	mēdèn hólōs sỳ lypoû	Have no grief at all
πρὸς ὀλίγον ἔστι τὸ ζῆν	pròs olígon ésti tò zên	Life is only a short while
τὸ τέλος ὁ χρόνος ἀπαιτεῖ.	tò télos ho khrónos apaiteî.	And time demands its toll.

The tune is joyful, a celebration of Euterpe's life. The song is known as the Seikilos epitaph.

· Centuries later in Vienna, Beethoven too sought to convey the feeling of joy in perhaps history's best-loved and most performed symphony movement, *Ode to Joy* within the Ninth Symphony. As a powerful means of expressing and stirring emotions, music has always played an important role in human lives, uplifting spirits throughout the generations. Humanity has continuously created new musical techniques and styles. But Vienna's cultural achievement was significant. By producing so many musical compositions that revolutionized the art form and continue to resonate with audiences centuries later, Vienna earned its moniker, the City of Music.

The musical legacy of Vienna has enriched humanity. The city also demonstrated the role of prosperity in funding great works of art. Vienna dramatically changed how music is performed, gave the world more groundbreaking composers than any other city, and was the birthplace of compositions that many believe represent the pinnacle of musical achievement. Vienna has therefore earned its place as a center of progress.

30
Manchester
INDUSTRIALIZATION

Our next center of progress is Manchester during the first Industrial Revolution (1760–1850). Sometimes called "the first industrial city," Manchester epitomized the rapid changes of an era that transformed human existence more than any other period in history. Manchester was among the earliest cities to experience industrialization. The city's metamorphosis wasn't easy, as it entailed working and living conditions far below those that we are used to today. But Manchester ultimately helped uplift humanity by paving the way to the postindustrial prosperity that so many of us now enjoy.

Today, Manchester is the fifth most populous city in the United Kingdom. The city is famous for its soccer team, Manchester United, which has won more trophies than any other English football club (i.e., soccer team). Nicknamed the Red Devils, Manchester's is among the world's most popular and highest-earning soccer teams. Manchester is also known for its large research university, where the atom was first split in 1917. The University of Manchester operates the Jodrell Bank Observatory, a designated UNESCO World Heritage site because of its substantive impact on research during the start of the Space Age. Manchester has also made notable contributions to music, producing groups such as the Bee Gees, who were among the best-selling musical artists in history. Much of the city's architecture dates to the industrial era, with many prominent warehouses, factories, railway viaducts, and canals still remaining.

The area where Manchester now stands has been inhabited since at least the Bronze Age, originally by ancient Celtic Britons. In about the 70s CE, Romans conquered the area. They called the outpost Mamucium. That is thought to be a Latinization of the prior name for the settlement in Old Brittonic, which likely meant "breast-shaped hill." Mamucium eventually became known as Manchester, with the Old English "chester" suffix coming from the Latin *castrum,* meaning "fortified town." After the Romans departed Britain, the settlement of Manchester changed hands among several kingdoms during the Middle Ages and the Norman Conquest. Manchester first became known for the cloth trade in the 14th century, when a wave of immigrant Flemish weavers who produced linen and wool settled in the town. By the 16th century, Manchester's economy revolved around the wool trade. A cottage industry, wool production was a slow and painstaking process that took place within individual households.

Manchester was a flourishing but small market town before the Industrial Revolution, with a population of fewer than 10,000 people at the start of the 18th century. As technological advances increased the efficiency of the cloth business, the city's growth began to take off in the 1760s. The city's canals, cotton-friendly climate, and location allowed for easy transport of goods into and out of the city, destining Manchester to become a key industrial center once the right technology arose.

The Industrial Revolution is often said to have begun when the spinning jenny was invented in Oswaldtwistle, 25 miles northwest of Manchester, in 1764 or 1765. The spinning jenny was a frame for spinning wool or cotton with increased speed by using multiple spindles. It represented the first fully mechanized production process. Then in 1771, another new invention, the water frame, was installed in a Cromford factory 50 miles southeast of Manchester. That invention used a waterwheel to power a spinning frame. Around 1779, in Bolton, which is located 15 miles northwest of Manchester, the inventor Samuel Crompton (1753–1827) combined aspects of the spinning jenny and the water frame into the "spinning mule."

The spinning mule greatly accelerated the process of producing yarn. In fact, versions of the spinning mule are still used today in the production of yarns from certain delicate fibers, such as alpaca hair. Water-powered textile mills making use of the new technology soon popped up across the region.

In 1781, just two years after the spinning mule's introduction, the development of viable steam engines then enabled the growth of larger, more powerful, steam-powered textile mills. Steam power was a game-changer. Although humanity had known about steam power since Hero of Alexandria (mentioned in Chapter 11) demonstrated the phenomenon as a novelty in the first century CE, finally gaining the ability to harness steam in a practical way was the pivotal moment of the Industrial Revolution. Improved steam engines led to the swift industrialization of England's cloth industry, allowing the spinning and weaving of textiles with a rapidity never before achieved.

Manchester opened its first cotton mill in 1782—the five-story Shudehill Mill, also sometimes called Simpson's Mill. It made use of a 30-foot waterwheel and cutting-edge steam power. By 1800, Manchester was described as "steam mill mad," with more than 40 mills. By that same year, the city's population had grown almost tenfold from the start of the 18th century, reaching about 89,000 souls. Between 1801 and the 1820s, the population doubled. By 1830, Manchester boasted 99 distinct cotton-spinning mills. Heavy industry was soon common in other areas of northern England as well, such as Liverpool and Birmingham.

That year, the world's first modern railway, the Liverpool and Manchester, opened and supercharged Manchester's already-booming textile industry. It did so by speeding up the import of raw materials from Liverpool's ports into Manchester's factories, as well as the export of finished textile products out of Manchester. The 31-mile-long Liverpool and Manchester Railway was both the first railway to exclusively serve steam-powered automotives and the world's first intercity railway. It was also the first railway to use a double track, to operate fully on a regular timetable, to use a signaling system, and to transport mail. By the end of the first Industrial Revolution in 1850, Manchester was home to some 400,000 people. The erstwhile obscure market town had become second only to London in importance within Britain and came to be called that nation's "second city."

The swelling of the population was driven by an inpouring of young men and women from the English countryside and from Ireland, drawn by the promise of work in the new factories and mills. Compared with backbreaking agricultural labor or lives of domestic servitude (in an era when many employers beat their servants with impunity), many people found even the famously harsh working conditions within the mills to be preferable to their other options. Mills paid high wages compared with opportunities in rural areas, and most migrants to the city saw an appreciable rise in their incomes. Gradually, and for the first time in history, a large middle class emerged.

That is not to make light of the working environment within early Industrial Revolution–era Manchester's factories, which saw long hours, high rates of injury, and the frequent use of child labor. Although it must be noted that child labor was not an innovation of the Industrial Revolution—it had tragically existed since time immemorial among the poor. In fact, it was only during the Industrial Revolution that living conditions improved so much that child labor began to come under scrutiny, resulting in Britain's 1833 Factory Act. That act is thought to be the world's first anti-child-labor legislation. Other acts followed.

If you could visit Manchester during the first Industrial Revolution, you would probably enter the city via a steam-powered locomotive, and your first sight of the city would be its bustling train station. You would emerge from the station into a city defined by a skyline of industrial smokestacks that the poet William Blake famously described as "dark Satanic mills." In 1814, the British civil servant Johann May described that skyline as a sign of technological progress:

> Manchester [has] hundreds of factories . . . which tower up to five and six storeys in height. Huge chimneys at the side of these buildings belch forth black coal vapours, and this tells us that powerful steam engines are used. The clouds of vapour can be viewed afar. The houses are blackened by it.

The sound might have been deafening. The French political philosopher Alexis de Tocqueville (1805–1859) described Manchester in 1835 by noting that the "crunching wheels of machinery, the shriek of steam from boilers, the regular beat of the looms . . . are the noises from which you can never escape." And among the people in the streets you might have observed various protestors. The city was at the vanguard of radical political movements ranging from women's suffrage and anti–Corn Law advocacy to communism. (The Corn Laws were tariffs placed on food and grain imported into Britain, and their repeal in 1846 thanks to the advocacy of reformers such as Richard Cobden [1804–1865] and John Bright [1811–1889] was a historic move toward free trade.)

The German political philosopher and Karl Marx's primary financial patron, Friedrich Engels (1820–1895), came to

Manchester in 1842. He worked there as a cotton merchant by day and opined about the state of the city's poor by night, culminating in the publication of *The Condition of the Working Class in England* in 1844. One passage on Manchester's slums reads:

> In a rather deep hole . . . surrounded on all four sides by tall factories . . . stand two groups of about 200 cottages, built chiefly back to back, in which live about 4,000 human beings, most of them Irish. The cottages are old, dirty, and of the smallest sort, the streets uneven, fallen into ruts and in part without drains or pavement; masses of refuse, offal, and sickening filth lie among standing pools in all directions.

What Engels failed to notice was that for the first time in history, such abject levels of poverty were actually in decline—within his own lifetime the average Englishman became three times richer.

Penury had always been the default state for the vast majority of humanity. Then, suddenly, average incomes not only began to rise, but rose exponentially. The famous hockey stick chart (see chart on next page), perhaps the most important graph in the world, illustrates the dramatic shift. Humanity has produced more economic output over the past two centuries than in all of the previous centuries combined. That explosion of wealth creation soon led to a massive decrease in the rate of poverty and to improvements in living standards. Not long after incomes took off, life expectancy followed. Economic historian Deirdre McCloskey calls the change the Great Enrichment.

Real gross domestic product per capita, 2011 U.S. dollars

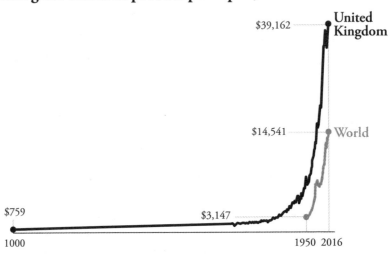

Engels lived in Manchester on and off for three decades. In Manchester, he was visited multiple times by his friend and fellow German philosopher Karl Marx (1818–1883). Moved by the state of the poor in Manchester and other factory cities, and failing to recognize the ongoing Great Enrichment, the two men developed a political philosophy that aimed to create a workers' paradise.

Their proposed solutions tragically led to far worse suffering—including food shortages, gulags, 100 million deaths, and psychological scars that still echo to this day, with heightened dishonesty and lower trust persisting in formerly communist areas. Ironically, Marx and Engels's goals of shorter workdays and higher incomes have been achieved within a market economy.

As the quintessential industrial city, there is no doubt that Manchester earned its nickname, Workshop of the World. As a key early center of industrialization, Manchester underwent an at times difficult transition with profound effects. The unprecedented prosperity created by industrialization eventually allowed for the improvement of working conditions and the heightened living standards that characterize postindustrial affluence. For helping weave the fabric of the modern world, Manchester is deservedly a center of progress.

31

London

EMANCIPATION

Our next center of progress is London during the late 18th and early 19th centuries, when the city played host to debates on the nature of human rights that would change the world. Today, we take for granted the norm that no person can buy or sell another human being, but it took humanity a long time to arrive at that norm. Slavery was accepted and rarely questioned for millennia throughout the world, but today, slavery is illegal in all countries. Although the abhorrent practice continues illegally in some parts of the world, the banning of slavery is nonetheless worth celebrating. Legal battles fought in London and legislative actions taken in London helped end the global slave trade and bring about the dramatic change in attitudes about slavery—an invaluable victory for human freedom.

Today, London is a city that needs no introduction. It is well-known as one of the world's foremost global cities, as well as the capital and most populous city in the United Kingdom. London is recognized as a center of commerce, finance, the arts, education, and research, and is among the globe's most popular tourist destinations. It is home to Buckingham Palace, the iconic Big Ben clock tower, the British Museum, and Europe's tallest Ferris wheel—the London Eye. It also houses four different UNESCO World Heritage sites: Westminster Abbey, the medieval Tower of London, Kew Gardens, and Maritime Greenwich.

Evidence suggests that the site of present-day London has been inhabited since at least the Bronze Age. However, the site's importance began when the Romans founded a port settlement there in 43 CE. It was known as Londinium, which soon became a regional trade hub, major road nexus, and the capital of Roman Britain during most of the time that the Romans ruled the province of Britannia. Once the Romans left Britain, Anglo-Saxons gained rule in London and the city became the capital of the eventual Kingdom of England. After the Norman Conquest in 1066, William the Conqueror (c. 1028–1087) became England's king, and it was during his rule that London was first linked to attempts to limit slavery.

In different parts of the world, slavery had long been subject to sporadic criticism, various limits, and even brief bans. For example, Emperor Wang Mang banned slavery in China in 9 CE. It was reinstated soon afterward. In the seventh century, the Frankish queen Balthild, herself a former slave, helped enact reforms that prevented the trade of Christian slaves. In the 740s,

Pope Zachary banned the sale of Christian slaves to Muslims. And in 873, Pope John VIII similarly called the enslavement of Christians sinful and advocated for the slaves' release.

But the early attempt to restrict slavery that would have the most lasting impact occurred in London. According to the Domesday Book, an extensive survey of England and parts of Wales completed in the 1080s, about 10 percent of people in the area were slaves. In 1080, William the Conqueror banned the sale of slaves to non-Christians. In 1102, the ecclesiastical Council of London banned the slave trade within England, decreeing, "Let no one dare hereafter to engage in the infamous business . . . of selling men like animals."

Within a generation, slavery had all but vanished in England. It was replaced by serfdom. Unlike slaves, serfs could at least own property. Also they were not at risk of separation from their families. Alas, they could not move, since they were perpetually confined to the land they worked. A feudal lord could sell that land, thus changing whom the serf served, but serfs themselves were not sold. (It is important to note that despite banning slavery, medieval England would not meet modern standards of human rights. Consider religious freedom. In 1290, the land's Jews were forcibly expelled during the reign of Edward I.)

Since time immemorial, every major civilization practiced some form of slavery for most of history. Slavery has existed since at least 3500 BCE, when the ancient Sumerians practiced it. Improvements in seafaring led to globalization of the slave trade. The Atlantic slave trade, for example, lasted from the 16th to the 19th century and involved the transport of millions of sub-Saharan Africans across the ocean to live in bondage.

The first foreign slave traders in sub-Saharan Africa were Arab (Saudi Arabia, in fact, did not outlaw slavery until 1962). However, Europeans were soon prominent participants in the maritime slave trade, transporting roughly 11 million slaves out of Africa. The first and the worst offender was Portugal, which transported about 5 million slaves from the African slave markets mainly to its colony of Brazil.

Britain transported the second-highest number of enslaved Africans (2.6 million) to its various colonies. At least 300,000 African slaves were shipped to Britain's North American colonies that would later become the United States. However, the near-total absence of slavery within Britain itself, which had persisted since the reforms of William the Conqueror, would prove critical to turning British hearts and minds against the institution.

As is widely known, the African slaves were treated as chattel rather than as people, and the conditions of the slave ships were horrific, with many enslaved people not surviving the journey. Most of those who made it through the voyage then lived the nightmare of forced, grueling agricultural labor on New World plantations. Although the brutal treatment of slaves in North America was abominable, slaves on the Caribbean and Brazilian plantations endured the worst conditions and suffered the highest fatality rates.

An enslaved Barbadian teenager, Jonathan Strong, was brought to London by his slave master, who in 1765 beat Strong with a pistol and left him for dead in the street. Strong, bleeding and left mostly blind by the attack, ended up at a medical clinic for the poor held in Mincing Lane. There, as he received treatment for his wounds from a physician, Strong made an impression on the physician's visiting brother—Granville Sharp (1735–1813).

Sharp, who was born in Durham but had lived in London since the age of 15, was forever changed by the encounter. He and his brother took Strong to a hospital and paid for the latter's months-long treatment there. But soon after becoming well enough to leave the hospital, Strong was recaptured by his former enslaver, who attempted to sell Strong to a Jamaican plantation.

Sharp successfully defended Strong's freedom, defeating Strong's former slave master in court—but only on a technicality. Tragically, Strong's health was permanently damaged from the pistol attack and he passed away at the age of 25 in 1770. Sharp devoted himself to bringing about a definitive legal ruling on the question of whether a man could be compelled to leave Britain and enter slavery, and his efforts earned him a reputation as an Enlightenment thinker and anti-slavery campaigner. He was not alone. The abolition movement in Britain was growing.

In 1769, another slaver from the colonies attempted to bring an enslaved man, James Somerset, to London. In 1771, Somerset escaped. In less than two months, Somerset was captured and arrangements were made to sell him again into slavery in Jamaica. Three Londoners applied for Somerset to receive a hearing and their petition was granted. Many concerned Britons sent money to launch a legal defense for Somerset, but several lawyers volunteered to do the case pro bono. Sharp advised Somerset's lawyers extensively.

One barrister, William Davy, famously cited in Somerset's defense an alleged 1569 case in which a cartwright attempted to bring a slave to England from Russia. In that case, it was resolved that England's air was "too pure" for a slave to breathe and that anyone in England was therefore free. Or as the London-born jurist Sir William Blackstone (1723–1780) had once put it, "The spirit of liberty is so deeply ingrained in our constitution that a slave, the moment he lands in England, is free."

Somerset won his case. The ruling stated that, while in Britain, Somerset was free. Furthermore, he could not be forced to depart the country. The ruling was a turning point.

By the time of the Somerset judgment, the absence of slavery in Britain had become a matter of British pride. It was also a moral issue among several Enlightenment thinkers; members of the clergy, including Anglican cleric John Newton (1725–1807) himself both a former slave held in what is now Sierra Leone as well as a reformed ex-slave ship captain, and the writer of the well-loved hymn "Amazing Grace"; and the general public.

By 1807, thanks to mounting public pressure and the work of tireless reformers such as William Wilberforce (1759–1833) in Britain's London-based Parliament in Westminster, Britain banned the international slave trade with the Slave Trade Act. When diplomatic efforts to pressure Paris and Vienna to sign similar legislation proved futile, public support for the use of force rose.

Decisionmakers in London ordered the Royal Navy to form the West Africa Squadron in 1808 to blockade West Africa and stop the movement of slave-transporting ships across the Atlantic Ocean. By the 1850s, the West Africa Squadron consisted of approximately 25 ships, 2,000 British men, and 1,000 additional crew members who were recruited locally, mainly from what is now Liberia. The British naval officers were paid a reward for each slave they freed, but the main incentive for many was humanitarian—by that point, anti-slavery efforts were hugely popular in Britain. As the poet Alfred Tennyson (1809–1892) put it, "This spirit of chivalry . . . we see it in acts of heroism by land and sea, in fights against the slave trade."

Between 1808 and 1860, the West Africa Squadron successfully hunted down at least 1,600 slave ships and freed about 150,000 African slaves. Spain and Portugal attempted to continue the slave trade, often purchasing slaves from African sellers. In the mid-18th century, King Tegbesu of Dahomey (reigned 1740–1774) in present-day Benin drew the equivalent of about 250,000 pounds annually—the greatest part of his income—selling slaves captured in battle to Europeans. His successor to the throne, Ghezo, (reigned 1818–1859) declared in 1840 in response to British pressures to stop selling slaves: "The slave trade is the ruling principle of my people. It is the source and the glory of their wealth . . . the mother lulls the child to sleep with notes of triumph over an enemy reduced to slavery." His acceptance of slavery demonstrates how deeply the practice was still ingrained at the time, across the globe.

The British navy eventually blockaded Brazil as well and succeeded in halting the Brazilian slave trade in 1852. (Tragically, ownership of, rather than trade of, slaves remained legal there until 1888.) But the effects of the abolition movement that started in London did not stop there. The movement saw a revival in the 1860s, when David Livingstone (1813–1873), the Scottish physician and prominent member of the London

Missionary Society, published reports describing the Arab slave trade in Africa that too moved the British public. In the 1870s, the British navy again devoted resources to stopping the slave trade—this time by traders based in Zanzibar. Thanks in part to efforts launched in London, the number of countries with legal slavery plummeted throughout the 19th century.

Although the lawmakers of London during the 18th and 19th centuries were far from perfect, their anti-slavery zeal helped change the world for the better. As the Irish historian William Lecky (1838–1903) put it, "The unweary, unostentatious, and inglorious crusade of England against slavery may probably be regarded as among the three or four perfectly virtuous pages comprised in the history of nations."

It was in London that British abolitionists organized, won court and legislative victories, launched naval ships with the mission of emancipating slaves, and ultimately helped alter moral norms that had persisted since the dawn of civilization. For its critical role in ending the slave trade and denormalizing the institution of slavery, London is justly a center of progress.

32
Wellington
SUFFRAGE

Our next center of progress is Wellington during the late 19th century, when the city made New Zealand the first country in the world to grant women the right to vote. At the time, that was considered a radical move. The reformers who successfully petitioned New Zealand's Parliament then traveled the world, organizing suffrage movements in other countries. Today, thanks to the trend that began in Wellington, women can vote in every democracy, except the Vatican, where only cardinals vote in the papal conclave.

Today, Wellington is best known as the capital city of New Zealand and the southernmost capital in the world. The windy bayside city has a population of just over 200,000 people and a reputation for trendy shops and cafés, seafood, quirky bars, and craft breweries. It has quaint red cable cars and its historic Old Government Building, constructed in 1876, remains one of the world's largest wooden structures. Wellington is also home to Mount Victoria, the Te Papa Museum, and a wharf with frequent pop-up markets and art fairs. Young and entrepreneurial, Wellington has been ranked as one of the easiest places in the world to start a new business. It is also a creative arts and technology center, famed for the nearby Weta Studios' work on the *Lord of the Rings* movie franchise.

According to legend, the site where Wellington now stands was first discovered by the legendary Māori chief Kupe in the late 10th century. Over the following centuries, different Māori tribes settled in the area. The Māori called the area Te Whanganui-a-Tara, meaning "the great harbor of Tara," named after the man said to have first scouted the area on behalf of his father, Whātonga the Explorer. An alternative name was Te Upoko-o-te-Ika-a-Māui, meaning "the head of the fish of Māui," referencing the mythical demi-god Māui who caught a giant fish that transformed into the islands of New Zealand.

Noting the site's perfect location for trade, an English colonel purchased local land in 1839 from the Māori for British settlers. A business district soon blossomed around the harbor, transforming it into a busy port. The following year, representatives of the United Kingdom and various Māori chiefs signed the Treaty of Waitangi, which brought New Zealand into the British Empire and made the Māori British subjects. Wellington was the first major European settlement in New Zealand; it was named after Arthur Wellesley, first duke of Wellington (1769–1852)—one of many tributes to the famed prime minister and military leader who defeated Napoleon at the Battle of Waterloo in 1815.

Interestingly, New Zealand has no widely agreed upon "Independence Day." Rather, the country's sovereignty seems to have come about gradually, with key events in 1857, 1907, 1947, and 1987. It was not until that last year that New Zealand "unilaterally revoked all residual United Kingdom legislative power" over the nation, according to the New Zealand Parliament's official website.

The colonial nation's demographics changed rapidly. By 1886, the majority of non-Māori residents were New Zealand–born rather than British-born immigrants, although the latter continued to stream into the country. Although many people thought of themselves as British, the term "New Zealander" was becoming more common. By 1896, New Zealand was home to more than 700,000 British immigrants and their descendants, as well as close to 40,000 Māori people, who had lost large stretches of territory to colonial forces during the violent New Zealand Wars (1845–1872).

Throughout most of history, women were largely excluded from politics, though it is important to remember most men were excluded as well. Political power tended to be concentrated among a small group, such as a royal family, while the majority of people, both male and female, lacked any meaningful say in political decisions. However, although history has certainly had its share of politically powerful women, from the Byzantine empress Theodora (c. 497–548) to the Chinese empress Wu Zetian (624–705), the majority of rulers in all major civilizations have been male.

In other words, in a world with highly exclusionary political institutions that left almost everyone out, women were even more likely to be left out than men. Likewise, when a wave of democratization expanded the pool of political participation to an unprecedented share of the population in the 19th century, the voting rolls still excluded women.

Young New Zealand was no exception, and women were initially denied the right to vote. A popular belief was that women were suited only to the domestic sphere, leaving "public life" to the men. But by the late 19th century, as more women entered professional fields previously staffed solely by men, women began to be viewed as more capable of participating in the public sphere.

These changes helped galvanize the suffrage movement in New Zealand. Suffragettes such as Kate Sheppard (1847–1934) gathered signatures to provide evidence of growing public support for female suffrage. In 1891, 1892, and 1893, the suffragettes compiled a series of massive petitions calling on Parliament to enact female suffrage. The 1893 petition for female suffrage gained some 24,000 signatures, and the sheets of paper, when glued together, formed a 270-meter roll, which was then submitted to the Parliament in Wellington.

The suffrage movement was aided by widespread support from New Zealand's men. As a "colonial frontier" country, New Zealand had far more men than women. That happened because single men were generally more likely to immigrate abroad to harsh environs. Desperate for companionship, the country's men sought to attract more women to New Zealand and often romanticized the latter. Many New Zealanders believed that an influx of women would exert a stabilizing effect on society, lowering crime rates, decreasing rates of alcohol use, and improving morality.

Indeed, research suggests that highly unequal sex ratios can cause problems: societies with far fewer women than men see higher rates of male depression, aggression, and violent crime. It is most likely that those negative effects stem from tensions that arise when a large number of men in a society lack companionship and feel that they have little hope of ever finding a wife.

However, the popular view in 19th-century New Zealand was that women were in some ways morally superior to men, or more likely to act for the good of society. Building on that belief, suffrage supporters cast women as "moral citizens" and argued that a society where women could vote would become more virtuous. In particular, the women's suffrage movement was closely connected to the alcohol prohibition movement. Men who supported alcohol prohibition on moral grounds were thus highly likely to support giving women the right to vote.

New Zealand was not an outlier—the other places that granted women the right to vote early on were also typically "frontier" societies. Like New Zealand, those places had a large male majority and were motivated by a belief that female voters were morality minded and would rally against social ills. The most prominent of those perceived ills were alcohol and, in the western United States, polygamy—as practiced by some adherents of the young Latter-Day Saints movement. (Prejudice against that group was widespread at the time.) It was also believed that women would oppose unnecessary wars and promote a more dovish foreign policy. Among the earliest adopters of female suffrage in the United States were the frontier western mountain states Wyoming (in 1869, pre-statehood), Utah (1870, also pre-statehood), Colorado (1893), and Idaho (1895).

The frontier territories of South Australia (1894) and Western Australia (1899) followed the same pattern.

But New Zealand led the way as the first country to give women the right to vote. Moved by the suffragettes' tireless efforts and their numerous male allies, the government embarked on a radical experiment. In Wellington, the governing Lord Glasgow signed a new Electoral Act into law on September 19, 1893. The act gave women the right to vote in parliamentary elections.

Ever since then, women have taken an active role in governing the country from the capital of Wellington. New Zealand has not only had three different female prime ministers, but also has had women in each of its key constitutional positions in government. At various times, New Zealand has had a female prime minister, governor general, speaker of the House of Representatives, attorney general, and chief justice. The country remains proud of the pioneering step toward legal gender equality enacted in Wellington, even featuring suffragette leader Sheppard on the $10 banknote.

After her legislative victory, Sheppard and her allies toured several other countries and helped organize suffrage movements abroad.

Although women's voting and running for office may seem commonplace now, it was revolutionary at the time. For perspective, women in parts of the United States could not vote until 1920, and the United Kingdom did not grant women full, equal voting rights until 1928. Spain only granted women universal suffrage in 1931. France did so in 1945. Switzerland waited until 1971. Liechtenstein held out until 1984. And Saudi Arabia refused to budge until 2015. Today, women can vote almost everywhere.

As New Zealand's seat of government, Wellington was at the center of the first successful campaign to grant a country's women the right to vote. For playing host to a groundbreaking legislative victory for women's suffrage, Wellington is rightly a center of progress.

33

Chicago
RAILROADS

Our next center of progress is Chicago during the Age of Steam. Chicago played a central role in the popularization of rail transportation and remains the most important railroad center in North America today.

With about 2.7 million inhabitants, Chicago is the third most populous city in the United States. It is a major hub of commerce, boasting a diverse economy. As the city that erected the first modern skyscraper in 1885, Chicago is well-known for its distinctive buildings and other contributions to architecture. For example, the so-called Windy City is home to the 1,451-foot-tall Willis Tower, previously called the Sears Tower.

That structure was the tallest building in the world for almost a quarter century. It is still the third-tallest building in the United States, and its observation deck serves as a tourist attraction.

The city is also famous for its music, food (such as the city's signature deep-dish pizza), arts scene, sports (particularly the storied Chicago Cubs baseball team), and research universities. Those include Northwestern University and the University of Chicago. The latter gave the world the influential Chicago schools of economics and of sociology. Chicago is a cultural mixing bowl with large Italian, Polish, and Irish American populations, among others. Every year, during the Saint Patrick's Day celebration, which honors the patron saint of Ireland, the Chicago River that flows through the city is dyed green.

Even putting railroads aside, Chicago is an important transportation center. The city's O'Hare International Airport ranked as the fourth busiest in the world in 2022. And the area surrounding Chicago has the largest number of federal highways in the United States.

The site where Chicago now stands was first inhabited by various native tribes. Chicago's attractive location between the Great Lakes and navigable Mississippi River waterways made it a transportation center even then. The first nonnative settlers of the area spoke French. The name "Chicago" comes from the French pronunciation of a word used by the local indigenous people for a kind of wild garlic that grew abundantly in the area. (In fact, the vegetable still abounds and can be found in many Chicago restaurant dishes and artisan grocery stores.)

The first nonindigenous Chicagoan was Jean Baptiste Point du Sable (c. before 1750–1818), a Haitian-born frontiersman of

African descent who married a native woman and settled in the area. He made a living as a trader and is widely considered to be "Chicago's founder." Du Sable's business flourished and made him a wealthy man. The small settlement he began at the mouth of the Chicago River would one day help enrich humanity.

Chicago was rural at first. The town was officially incorporated in 1837 with a modest population of just 350 residents. However, the settlement was surrounded by rich farmland and was well situated to transport food by boat throughout the Great Lakes region. As early as the 1830s, entrepreneurs saw Chicago's potential as a transportation hub and began buying land in a flurry of speculation. By 1840, the little boomtown boasted 4,000 inhabitants. By 1850, it had almost 30,000 people.

Then the trains started arriving, and the city was never the same. Chicago's inaugural railroad was the Galena and Chicago Union. It welcomed its first locomotive, the Pioneer, on October 10, 1848. Nearly overnight, the city became a major commercial center. In 1852, one Chicagoan asked, "Can it be wondered at, that our city doubles its population within three years; that men who were trading in small seven-by-nine tenements, now find splendid brick or marble stores scarcely large enough to accommodate their customers?"

A stunned visitor to Chicago during the 1850s, Sara Jane Lippincott (1823–1904), wrote, "The growth of this city is one of the most amazing things in the history of modern civilization." She referred to Chicago as "the lightning city." Starting in 1857, durable steel rails—still the standard around the globe—replaced cast-iron rails. That innovation allowed trains to move twice as fast as before, greatly improving trains' practicality and further boosting steam transportation.

Chicago's rapidly rising population brought new public health challenges. An insufficient waste drainage system allowed pathogens to infect the water supply and caused outbreaks of illnesses such as typhoid and dysentery. One 1854 bout of cholera killed 6 percent of the city's population. Recognizing the problem, private property owners and city leaders cooperated to improve the city's drainage system in the late 1850s and 1860s. To make room for new sewers, they lifted the city 14 feet, in a Herculean feat of engineering. The "Raising of Chicago," as the endeavor came to be known, was accomplished piecemeal by lifting the city's massive brick buildings, streets, and sidewalks using large jackscrews operated by hundreds of men. If that is difficult to imagine, see the illustration below.

It was perhaps the most striking event of the modern sanitation movement, which also later reversed the flow of the Chicago River for sanitary purposes. Nobel Prize–winning economist

The "Raising of Chicago"

Angus Deaton partly credits improvements in sanitation with the dramatic rise in human life expectancy.

By 1870, Chicago's population had grown to almost 300,000 souls. Then tragedy struck. On a series of dry October days in 1871, a fire swept through Chicago. The flames claimed some 300 lives, destroyed about 17,500 buildings, and left more than 100,000 Chicagoans (i.e., over one-third of the city's people) homeless. According to legend, the Great Chicago Fire was sparked by a lantern kicked over by a cow belonging to Catherine O'Leary (1827–1895), an Irish immigrant. The fire's true origin remains a mystery, but the tale of "Mrs. O'Leary's cow" has entered popular culture, appearing in numerous songs and films. Regrettably, the story was fueled by anti-Irish sentiment. Chicago's city council officially exonerated the O'Leary family and the infamous cow in 1997, to the relief of Mrs. O'Leary's great-great-grandchildren.

Chicago rose from the ashes like a mythic phoenix to make its greatest contributions to human progress. After the Great Chicago Fire, the city was rebuilt around the rail industry. Chicago's central location helped the city contribute to the meteoric rise of rail-based commercial transportation. Recognizing Chicago's prime location, most railroad companies building westward chose the city for their headquarters. The city thus also became a major manufacturing center for railroad equipment.

The roar of passenger and freight trains soon filled the air around the city's six bustling intercity terminals. Municipal and regional commuter trains also appeared and redefined intracity transport. Chicago's Union Station still looks as it did during the golden age of rail and is today the United States' third-busiest train station.

Recent research suggests that the development of a nation-wide transportation system, particularly railroads, helped the United States urbanize and industrialize in the 19th century. The "transportation revolution" made it easier for rural workers to relocate to urban locations and take up manufacturing work. Trains also let goods flow more quickly across the country, allowing for greater regional economic specialization. As the country's Northeast region industrialized, the Midwest earned its nickname, America's Breadbasket, by producing wheat to support the country's swelling population.

Freight trains loaded with goods from other cities arrived at the central yards of Chicago. There, workers classified the goods. They then transferred the arrivals to massive sorting yards on the city's outskirts. Several railroad companies also operated a meatpacking district known as the Union Yards, and Chicago soon counted among its nicknames Hog Butcher for the World as it supplied meat to the rest of the country, delivered by rail. Chicago's meatpackers are sometimes credited with inventing the modern "assembly line," using conveyor belts in a way that may have inspired the industrialist Henry Ford (1863–1947) to use a similar method in automobile production.

Chicago's meatpacking district provided the subject of the grim 1906 novel *The Jungle* by Upton Sinclair (1878–1968), an outspoken socialist who sought to bring attention to harsh labor conditions. The 1914 poem "Chicago" by Carl Sandburg (1878–1967) presents a more nuanced view of the lives of the city's working class. Sandburg acknowledges Chicago's rough reputation while celebrating the spirit of the city's people and noting the pride they took in their work. The poem reads, in part:

> I turn once more to those who sneer at this my city, and
> I give them back the sneer and say to them:
> Come and show me another city with lifted head
> singing so proud to be alive and coarse and strong and
> cunning.
> . . . .
> Laughing the stormy, husky, brawling laughter of
> Youth, half-naked, sweating, proud to be Hog Butcher,
> Tool Maker, Stacker of Wheat, Player with Railroads
> and Freight Handler to the Nation.

As Chicago prospered, the city became a center of culture and innovation, with particularly notable contributions to transportation technology. As host of the Chicago World's Fair in 1893, the city gave humanity several new inventions. Those included the Ferris wheel (also called the Chicago wheel), the moving walkway, and the first third rail.

By 1900, Chicago was the fifth most populous city in the world and the second most populous in the United States, after New York City. If you could visit Chicago during the Age of Steam, you would enter a city jam-packed with pedestrians, horse-drawn carts, streetcars, and, of course, trains. Around 2,000 trains, including freight trains, arrived and departed the city each day. Rail transport had come a long way from the days when people doubted whether steam locomotives could outrace horses.

Steam transportation helped create the modern world, and no city was more central to the so-called rail revolution than

Chicago. It was once commonly said that "all roads lead to Rome." That city's groundbreaking road system earned Rome its place as a center of progress. Today, it could as easily be said that "all railways lead to Chicago." For lending steam to urbanization, industrialization, and ultimately the Great Enrichment, Chicago during the golden age of train travel has arrived as a center of progress.

34

Los Angeles

CINEMA

Our next center of progress is Los Angeles during the golden age of Hollywood (1910s–1960s). The city pioneered new filmmaking styles that were soon adopted globally, giving the world some of its most iconic and beloved movies in the process. Los Angeles's Hollywood neighborhood is synonymous with filmmaking, representing the city's unparalleled cinematic contributions.

With some 4 million inhabitants, Los Angeles is the second most populous city in the United States. (And Los Angeles County has nearly 10 million residents.) However, it may well be the most glamorous, with many celebrities and movie stars calling Los Angeles home. The city is also known for its impressive sports centers and music venues, shopping and nightlife, pleasant Mediterranean climate, terrible traffic, beautiful beaches, and easygoing atmosphere. Two famous landmarks include Disneyland and Universal Studios Hollywood, film-related theme parks that attract about 18 million and 9 million annual visitors, respectively.

The site where Los Angeles now stands was first inhabited by native tribes, including the Chumash and Tongva. The first European explorer to discover the area was Juan Rodríguez Cabrillo (c. 1498–1543), who arrived in 1542. Los Angeles's Cabrillo Beach still bears his name. Spanish settlers founded a small ranching community at the site in 1781, calling it El Pueblo de Nuestra Señora la Reina de los Ángeles, meaning "the Town of Our Lady the Queen of the Angels." The name was soon shortened to Pueblo de los Ángeles.

The Mexican War of Independence saw control of the town pass from Spain to newly independent Mexico in 1821. Then, after the conclusion of the Mexican-American War (1846–1848), Mexico ceded the land comprising the future state of California to the United States.

That same year, gold was discovered in California. Hopeful miners poured into the area, and when California gained statehood in 1850, the migration intensified. True to its ranching roots, Los Angeles soon boasted the largest cattle herds in the state. The town gained a reputation as Queen of the Cow Counties for supplying beef and dairy products to feed the growing population of gold miners in the north.

Although most of Los Angeles County was cattle-ranch land, there were also a number of farms devoted to growing vegetables and citrus fruits. (To this day, the Los Angeles area remains a top producer of the nation's broccoli, spinach, tomatoes, and avocados.) As the local food industry prospered, the city proper began to grow, from about 1,600 inhabitants in 1850 to almost 6,000 people by 1870. Los Angeles wasn't anywhere close to even counting among the 100 largest cities in the United States at the time. (In fact, the only California city that made it into the top 100 in 1870 was Sacramento, at number 89.)

In 1883, a politician and real estate developer named Harvey Wilcox (1832–1891) and his significantly younger second wife, Daeida (1861–1914), moved to town. The pair wanted to try their hand at ranching and bought more than 100 acres of apricot and fig groves. When their ranch failed, they used the land to build a community of upscale homes. They named the new subdivision "Hollywood."

One story claims that Daeida was inspired by an estate with the same name in Illinois or by a town of the same name in Ohio. Others theorize that the Wilcoxes drew inspiration from a native shrub with red berries called toyon, or "California holly," which grows abundantly in the area. In tribute to that theory, the Los Angeles City Council named toyon the city's "official native plant" in 2012. Although the true origin of the name "Hollywood" remains disputed, Daeida has been nicknamed the Mother of Hollywood for her role in the story. (Ironically, she envisioned Hollywood as a Christian "temperance community" free of alcohol, gambling, and the like.)

In any case, Hollywood started as a small but wealthy enclave that by 1900 boasted a post office, a hotel, a livery stable, and even a streetcar. A banker and real estate magnate named

H. J. Whitley (1847–1931) moved into the subdivision in 1902. He further developed the area, building more luxury homes and bringing electricity, gas, and telephone lines to town. He has been nicknamed the Father of Hollywood.

Hollywood was officially incorporated in 1903. Unable to handle its sewage and water needs independently, Hollywood merged with the city of Los Angeles in 1910. By then, Los Angeles had about 300,000 people. That would top 1 million by 1930 and would grow to 2.5 million by 1960.

The city's explosive growth can be traced to one industry: filmmaking.

The first film to be completed in Hollywood was *The Count of Monte Cristo* in 1908. The medium of film was still young, and *The Count of Monte Cristo* was one of the first films to convey a fictional narrative. Filming began in our previous center of progress, Chicago, but by wrapping up production in Los Angeles, the film crew made history. Two years later came the first film produced start to finish in Hollywood, called *In Old California*. The first Los Angeles film studio appeared on Sunset Boulevard in 1911. Others followed suit, and what began as a trickle soon turned into a flood.

What led so many filmmakers to relocate to Los Angeles? The climate allowed outdoor filming year-round, the terrain was varied enough to provide a multitude of settings, the land and labor were cheap—and, most importantly, it was far away from the state of New Jersey, where the prolific inventor Thomas Edison (1847–1931) lived.

With exclusive control of many of the technologies needed to make films and operate movie theaters, Edison's Motion

Picture Patent Company had secured a near monoply on the industry. Edison held over 1,000 different patents and was notoriously litigious. Moreover, Edison's company was infamous for employing mobsters to extort and punish those who violated his film-related patents.

California was the perfect place to flee from Edison's wrath. Not only was it far from the East Coast mafia, but many California judges were hesitant to enforce Edison's intellectual property claims.

The Supreme Court eventually weighed in, ruling in 1915 that Edison's company had engaged in illegal anti-competitive behavior that was strangling the film industry. But by then, and certainly by the time that Edison's film-related patents had all expired, the cinema industry was already firmly planted in California. Edison has been called "the unintentional founder of Hollywood" for his role in driving the country's filmmakers to the West Coast.

Hollywood became the world leader in narrative silent films and continued to lead after the commercialization of "talkies," or films with sound, in the mid- to late 1920s. At first, such films were exclusively shorts. Then in 1927, Hollywood produced *The Jazz Singer*, the first feature-length movie to include the actors' voices. It was a hit. More and more aspiring actors and film producers flocked to Los Angeles to join the burgeoning industry.

In the 1930s, Los Angeles studios competed to wow audiences with innovative films. The Academy Awards, or Oscars, were first presented at a private dinner in a Los Angeles hotel in 1929 and first broadcast via radio in 1930. They remain the most prestigious awards in the entertainment industry to this day. Distinct movie genres soon emerged, including roman-tic comedies (including the beloved 1934 film *It Happened One Night*, which swept the Oscars and boasts a near-perfect score on the film review-aggregation website Rotten Tomatoes), musicals, westerns, and horror films, among others.

The innovations of that era continue to influence movies today. *King Kong* premiered in 1933. In 2021, its namesake giant ape appeared in his 12th feature film, this time battling Godzilla. Hollywood gave the world its first full-length an-imated feature film in 1937 with Walt Disney's (1901–1966) *Snow White and the Seven Dwarfs*. In 1939, Hollywood popu-larized color productions with the release of *The Wizard of Oz*. Although it was not the first color film, it was among the most influential in promoting the technology's widespread adoption. Major Hollywood studios still bear the names of successful early producers, such as Louis B. Mayer (c. 1884–1957), Samuel Goldwyn (c. 1879–1974), and the Warner brothers—Harry (1881–1958), Albert (1884–1967), Sam (1885–1927), and Jack (1892–1978).

In the 1940s, the iconic Hollywood sign first appeared in its current incarnation, replacing a sign reading Hollywoodland, erected in 1923. The next few decades saw the production of some of history's best-loved classic films. Those include *Citizen Kane* (1941); *Casablanca* (1942); *It's a Wonderful Life* (1946); *Singin' in the Rain* (1952); *Rear Window* (1954); *12 Angry Men* (1957); *Vertigo* (1952); *Psycho* (1960); *Breakfast at Tiffany's* (1961); and *The Good, the Bad and the Ugly* (1966). Some were partially filmed elsewhere; however, all were Hollywood creations—*Breakfast at Tiffany's*, for example, was set in New York but produced by a Hollywood studio, and many of its interior scenes were shot on a Hollywood set. Many of these films remain top-rated productions, beating decades of more recent movies to appear in the Internet Movie Database's top 100 films sorted by user rating.

As it transformed from a humble cattle town into the geographic center of filmmaking, Los Angeles came to define a new art form. Movies enrich humanity by providing entertainment, inspiration, laughter, and thrills. Moreover, films create cultural experiences that can bring people together, act as an artistic outlet, and even shift worldviews. Hollywood created modern cinema. Thus, every person who has ever enjoyed a movie, even one produced elsewhere, owes a debt of gratitude to Los Angeles. It is for those reasons that Los Angeles is a center of progress.

35
New York
FINANCE

As many great cities lay in ruins after World War II, New York City assumed a new global prominence and even overtook London's central position in the international financial markets. It soon became home to the world's largest and most prestigious stock market on Wall Street and forever changed finance. Wall Street is often considered to be both a symbol and a geographic center of capitalism.

Today, New York City is the most populous city in the United States, with more than 8 million people, and the greater New York metropolitan area, with over 20 million people, is among the world's most populous megacities.

In the American psyche, New York represents opportunity. Ellis Island was the historical gateway through which many immigrants arrived in the country during the 19th and 20th centuries, and New York remains a popular immigrant destination in the United States. In fact, it may be the world's most linguistically diverse city, with hundreds of languages spoken within its boundaries.

New York is also where ambitious Americans of all stripes traditionally go to make a name for themselves in industries as diverse as writing, theater, commerce, fashion, mass media, investment banking, and more. And those who make it often stick around. New York is home to more billionaire residents than any other city. The metropolis's nicknames include the City That Never Sleeps, the Big Apple, Gotham, the Capital of the World (popularized by *Charlotte's Web* author E. B. White), the Greatest City in the World, and, in the surrounding region, simply the City. Although sometimes that last moniker specifically refers to Manhattan. New York is made up of five boroughs: the Bronx, Brooklyn (once an independent city), Manhattan, Queens, and Staten Island.

New York's cultural and economic importance is difficult to overstate. The city is a popular tourist spot, home to the iconic Statue of Liberty, the towering Empire State Building, the famous Broadway theater district, the verdant Central Park, and bustling Times Square that is the site of the famous New Year's Eve ball drop. As such, New York has been called the world's "most photographed" city. It has been estimated that if the New York metropolitan area were a country, it would boast the eighth-largest economy in the world (a rank currently held by Italy). The city is also a research hub, home to more than 100 colleges and universities, including New York University, Columbia University, and The Rockefeller University.

Perhaps the city's geography destined it to be a center of commerce. Located in one of the world's largest natural harbors, the site where New York now stands was a logical place for human settlement. Originally, the area was inhabited by the Lenape people and other native American tribes. They used the natural waterways for fishing and to trade and wage war with nearby tribes. The first European to visit the site was an Italian, Giovanni da Verrazzano (1485–1528), who explored the region in service to the French, in 1524. He named the area New Angoulême, after the French king Francis I (who was known as Francis of Angoulême before assuming the French throne) and soon departed.

Then in 1609, the English explorer Henry Hudson (1565– disappeared 1611), the namesake of Hudson Bay and the Hudson River, arrived. He also soon left, but not before noting the large local beaver population. Beaver pelts were a valuable commodity. Word of Hudson's discovery spread quickly and inspired the Dutch to found several fur-trading outposts in the area in the early 17th century. Those included a 1624 settlement in what is now Manhattan, established by the Dutch West India Company. By 1626, the Dutch had constructed Fort Amsterdam, which would serve as the town's nucleus until the fort's demolition in 1790. The town was appropriately named New Amsterdam and served as the capital of the local Dutch colonies collectively called New Netherland. To this day, several local place names maintain Dutch origins, including Harlem and Brooklyn (named after Breukelen, a town in the Netherlands).

The Second Anglo-Dutch War (1665–1667), despite ending in a Dutch victory, resulted in the British gaining control of the city as part of a treaty. In exchange, the British ceded the Dutch what is now Suriname, as well as Run, a small island that produces nutmeg, in what is today Indonesia. At the time, it seemed as though the Dutch had gotten a far better deal than the British—nutmeg was extremely valuable, and the island complex that includes Run was famous in Europe, whereas New Amsterdam was a relatively obscure outpost. "Few would have believed a small trading village on the island of Manhattan was destined to become the modern metropolis of New York," according to Australian historian Ian Burnet.

After the exchange, New Amsterdam was promptly renamed New York after the English king Charles's brother James. James's title was Duke of York, and he was the admiral who led the campaign to conquer the city during the war. The city rapidly grew. By 1700, New York had a population of almost 5,000 people. By the time of American independence in 1776, New York's population was about 25,000. In 1800, New York City had approximately 60,000 inhabitants. Boosted by immigration, it had well over 3 million by 1900.

New York City took on its central importance in the post–World War II period. The Germans never acted on plans to bomb New York, deeming what they called the "Amerikabomber" operation too expensive. Thus, spared by the protective breadth of the Atlantic Ocean, New York emerged from the war not only unscathed but prospering and poised to dominate global business and culture.

By the late 1940s, New York had risen to become the world's biggest manufacturing center, boasting 40,000 factories, 1 million factory workers, and the world's busiest port, which handled 150 million tons of waterborne freight goods a year. New York was suddenly the city of choice for many top corporations doing business internationally—including Standard Oil, General Electric, and IBM. The nickname Headquarters City was added to the metropolis's collection of monikers. Even the newly formed United Nations was headquartered in New York (built 1947–1952). "The New York [of] 40 years ago was an American city," reminisced the British writer J. B. Priestley in 1947, "but today's glittering cosmopolis belongs to the world, if the world does not belong to it."

The city inherited Paris's role as the center of the art and fashion world. New York was a refuge for foreign artists fleeing war-battered Europe, like the Dutch painter Piet Mondrian (1872–1944), and a hothouse of creativity cultivating groundbreaking American artists like Jackson Pollock (1912–1956). The city's musical influence also expanded rapidly, from influential interpretations of classical music by the New York Philharmonic at Carnegie Hall to bebop, the new form of music pioneered in Harlem's nightclubs that would take the world by storm.

Above all, the city was at the center of postwar globalization. The British writer Beverley Nichols described the state of the megalopolis in 1948:

> There was the sense of New York as a great international city to which all the ends of the world had come. London used to be like that, but somehow one had forgotten it, so long had it been since the Hispanos and Isottas [luxury cars from Spain and Italy, respectively] had glided down Piccadilly, so many aeons since the tropical fruit had glowed in the Bond Street windows. Coming from that sort of London to America, in the old days, New York had seemed just—American; not typical of the continent, maybe, but American first and foremost. Now it was the center of the world.

Fittingly, the newly internationalized New York took on the role of the world's financial capital and the site of the world's two largest stock exchanges: the New York Stock Exchange and, later, Nasdaq (National Association of Securities Dealers Automated Quotations).

Since its humble origins in 1792, when 24 brokers signed the Buttonwood Agreement, thus establishing a securities trading operation in the city, the New York Stock Exchange has flourished in the face of adversity. The U.S. Civil War (1861–1865) helped the financial district expand by prompting more securities trading, and the stock exchange moved to its current location at 11 Wall Street in 1865. But it was World War II's aftermath that let the stock exchange gain unparalleled global prominence.

Credit cards were also among New York's postwar financial innovations. In 1946, a banker named John Biggins (c. 1910–1971) thought to create charge cards that could be used at various shops throughout Brooklyn. Shopkeepers could deposit the sales slips at Biggins's Flatbush National Bank, which would then bill cardholders.

In 1989, an iconic bronze statue known as the *Charging Bull* or the *Wall Street Bull* was erected in the Financial District of Manhattan to represent capitalism and prosperity. (A play on the term "bull market" that denotes positive market trends.)

As the symbol of capitalism, Wall Street became the target of the anti-capitalist Occupy Wall Street protest movement in 2011. The protestors were concerned about economic inequality, fearing that the prosperity created by the market system was not widely shared. In reality, shady types resembling the well-known movie villain and financier-caricature Gordon Gekko are hardly the only beneficiaries of financial markets. Wall Street plays an invaluable role in everything from facilitating ordinary Americans' retirements through their 401(k) accounts to funding promising innovations—ultimately expanding the economic pie and raising living standards. As my former colleague, securities lawyer Thaya Brook Knight, put it:

> At its core, here's what Wall Street does: It makes sure that companies doing useful things get the money they need to keep doing those things. Do you like your smartphone? Does it make your life easier? The company that made that phone got the money to develop the product and get it into the store where you bought it with the help of Wall Street. When a company wants to expand, or make a new product, or improve its old products, it needs money, and it often gets that money by selling stock or bonds. That helps those companies, the broader economy and consumers generally.

New York City remains the world's leading financial center and the heart of the U.S. finance industry, to the point that "Wall Street" has become shorthand for financial capitalism itself. Although many still consider Wall Street the world's financial center, new technologies have allowed investing to become increasingly decentralized. Today, anyone can buy and sell stocks using a smartphone while enjoying the comforts of home. Still, anyone who shares in the economic benefits of the financial sector should thank New York City for taking banking to new heights. It is appropriately a center of progress.

36
Hong Kong

NONINTERVENTIONISM

Our next center of progress is Hong Kong during its rapid free-market transformation in the 1960s. After a lengthy struggle with poverty, war, and disease, the city managed to rise to prosperity through classical liberal policies.

Today, the freedom that has been so key to Hong Kong's success is being stripped away. The Chinese mainland has cracked down on the city's political and civil liberties, leaving its future uncertain. But as my colleague Marian Tupy has noted, "No matter what lies ahead for Hong Kong, we should admire its rise to prosperity through liberal reforms."

The area where Hong Kong now stands has been inhabited since Paleolithic times, with some of the earliest residents being the She people. The small fishing village that would later become Hong Kong came under the rule of the Chinese Empire during the Qin dynasty (221–206 BCE). After the Mongol conquest in the 13th century, Hong Kong saw its first significant population increase as Song dynasty loyalists sought refuge in the obscure coastal outpost.

Hong Kong's position on the coast allowed its people to make a living by fishing, collecting salt, and diving for pearls. However, it also left them under the constant threat of bandits and pirates. One particularly notorious pirate was Cheung Po Tsai (1786–1822), said to have commanded a fleet of 600 pirate ships before the government recruited him to become a naval colonel and fight the Portuguese. His purported hideout on an island six miles off the coast of Hong Kong is now a tourist attraction.

China ceded much of Hong Kong to Britain in 1842 through the Treaty of Nanjing, which ended the First Opium War. As trade between China and Britain in silk, porcelain, and tea intensified, the port city became a transportation hub and grew quickly. That growth initially led to overcrowding and unsanitary conditions. Thus, it is unsurprising that when the third plague pandemic (1855–1945) took some 12 million lives globally and devastated Asia, it did not spare Hong Kong.

In 1894, the bubonic plague arrived in the city and killed over 93 percent of those infected. The plague and resulting exodus caused a major economic downturn, with 1,000 Hong Kongers departing daily at the pandemic's peak. In total, about 85,000 of the city's 200,000 ethnic Chinese residents left Hong Kong.

The bubonic plague remained endemic on the island until 1929. Even after the bubonic plague departed, Hong Kong remained unhygienic and ravaged by tuberculosis, or the "white plague."

Besides disease, life in Hong Kong was also complicated by war and instability on the Chinese mainland. In 1898, the Second Opium War (1898) brought Hong Kong's Kowloon Peninsula under British control.

The suffering in Hong Kong was well documented by journalist Martha Gellhorn (1908–1998), who arrived with her husband, the author Ernest Hemingway (1899–1961), in February 1941. Hemingway would later ironically refer to the trip as their honeymoon. Gellhorn wrote, "The streets were full of pavement sleepers at night. . . . The crimes were street vending without a license, and a fine no one could pay. These people were the real Hong Kong and this was the most cruel poverty, worse than any I had seen before." Yet things were about to get even worse for the city.

During the Second Sino-Japanese War (1937–1945), much of the material aid that China received from the Allied nations arrived through its ports—particularly the British colony of Hong Kong, which brought in roughly 40 percent of outside supplies. In other words, the city was a strategic target. British authorities evacuated European women and children from the city in anticipation of an attack. In December 1941, on the same morning that Japanese forces attacked Pearl Harbor in Hawaii, Japan also attacked Hong Kong, starting with an aerial bombardment. The British chose to blow up many of Hong Kong's bridges and other key points of infrastructure to slow the Japanese military's advance, but to no avail.

Following the Battle of Hong Kong, the Japanese occupied the city for three years and eight months (1941–1945). The Hong Kong University of Science and Technology refers to the episode as perhaps "the darkest period of Hong Kong's history." The occupying forces executed about 10,000 Hong Kong civilians and infamously tortured, raped, and mutilated many others. The situation prompted many Hong Kongers to flee, and the city's population rapidly shrank from 1.6 million to 600,000 people during the occupation. After the Japanese surrendered to American forces in 1945, the British returned to Hong Kong.

That same year, a 30-year-old civil servant from Scotland named Sir John James Cowperthwaite (1915–2006) came to the colony to help oversee its economic development as part of the Department of Supplies, Trade, and Industry. He was originally assigned to go to Hong Kong in 1941, but the Japanese occupation forced his reassignment to Sierra Leone. When he finally arrived in Hong Kong, he observed a war-ravaged city in an even worse state of poverty than Gellhorn had described. It was appropriately nicknamed the Barren Island. With the entrepôt business stalled, the British considered handing the seemingly hopeless city filled with war refugees back to China.

But Cowperthwaite had some ideas that would help transform Hong Kong from one of the poorest places on the planet to one of the most prosperous.

What was the miraculous intervention that he proposed? Simply allowing Hong Kong's people to rebuild their shops, engage in exchange, and ultimately save themselves and make their city rich. Cowperthwaite trusted in the capabilities of ordinary people to run their own lives and businesses. He and his fellow administrators provided the city freedom, public security, the rule of law, and a stable currency—and left the rest to the people. To put it simply, he enacted a policy of doing nothing. That isn't to say he actually did nothing; keeping the other bureaucrats in check kept him plenty busy. He would later claim one of the actions he was most proud of was to prevent collection of statistics that could potentially justify economic intervention.

Cowperthwaite rose steadily through the bureaucracy and eventually became Hong Kong's financial secretary, a post he occupied from 1961 to 1971. During the 1960s, many countries experimented with centralized economic planning and high degrees of public spending financed by heavy taxes and large deficits. The idea that governments should attempt to steer the economy, from industrial planning to intentional inflation, was virtually a global consensus. Cowperthwaite resisted the political pressure to follow suit. From 1964 to 1970, Britain was ruled by a Labour government that favored heavy-handed economic intervention, but Cowperthwaite ran constant interference to keep his compatriots from meddling with Hong Kong's market.

As the communist-controlled Chinese mainland violently purged any remnants of capitalism (among other things) during the reign of terror that was later called the Cultural Revolution (1966–1976), Hong Kong went down a markedly different path.

In 1961, in his first budget speech, Cowperthwaite opined, "In the long run, the aggregate of decisions of individual businessmen, exercising individual judgment in a free economy, even if often mistaken, is less likely to do harm than the centralized decisions of a government, and certainly the harm is likely to be counteracted faster."

He turned out to be right. Once freed, Hong Kong's economy became breathtakingly efficient and saw explosive economic growth. The city was among the first in East Asia to fully industrialize and just as rapidly moved to postindustrial prosperity. Hong Kong soon became an international center of finance and commerce, earning its nickname, Asia's World City. Hong Kong's economic rise dramatically improved the local standard of living. During Cowperthwaite's tenure as financial secretary, Hong Kong's real wages rose by 50 percent, and the number of households in acute poverty fell by two-thirds.

When the Scotsman arrived in Hong Kong in 1945, the average income in Hong Kong was less than 40 percent that of Great Britain. But by the time Hong Kong was returned to China in 1997, its average income was higher than Britain's (see the table below).

Cowperthwaite's successor, Sir Philip Haddon-Cave (1925–1999), named Cowperthwaite's strategy the "doctrine of

Real average income in Hong Kong and Great Britain, 1945–1997

	1945	1997	1945–1997
Hong Kong	$3,783	$35,327	$31,543 absolute change
United Kingdom	$9,567	$29,840	$20,273 absolute change
Hong Kong	40% relative to U.K.	118% relative to U.K.	834% increase
United Kingdom	253% relative to H.K.	84% relative to H.K.	212% increase

positive noninterventionism." Positive noninterventionism became the official policy of the Hong Kong government and remained so as recently as the 2010s. For years, the city boasted the world's freest economy, with bustling financial and trade industries and a human rights record far superior to that of the Chinese mainland.

Then in 2019, Beijing began requiring extradition of fugitives in Hong Kong to the mainland—eroding the independence of Hong Kong's legal system. In response to the resulting mass protests, the mainland Chinese government implemented a brutal crackdown on Hong Kong's political and economic independence. In July 2020, a new national security law imposed by the communist government in Beijing criminalized protests and stripped away several other freedoms previously enjoyed by Hong Kongers. More sweeping changes followed, such as an overhaul of Hong Kong's education system.

Hong Kong was returned to China on the condition that it would remain autonomous until 2047. But the "autonomous territory" is, sadly, no longer truly autonomous.

From a starving city plagued by war and poverty to a shining beacon of prosperity and freedom, Hong Kong's rise exemplified the potential of limited government, rule of law, economic freedom, and fiscal probity. Sadly, the pillars on which Hong Kong's success was built are now crumbling in the tightening fists of the Chinese Communist Party. Whatever the future may hold for the island city, its transformation reflects how much people can achieve when given the freedom to do so. This historic policy lesson merits Hong Kong's place as a center of progress.

37

Houston

SPACEFLIGHT

Our next center of progress is Houston during the 20th-century space race, the famous period of rivalry between the United States and the Soviet Union over which nation could achieve more in the realm of space exploration. Nicknamed Space City because it houses NASA's famed Mission Control Center, Houston has done more to advance space exploration than any other city.

Today, as the fourth most populous city in the United States (beaten only by our previous centers of progress New York, Los Angeles, and Chicago), Houston is a sprawling, busy port city. As the largest city in Texas and in the country's South,

Houston is also a thriving center of regional cultural traditions and boasts the world's largest livestock exhibition and rodeo. The Houston Rodeo draws millions of visitors annually and has attracted famous musical performers over the years ranging from Elvis Presley to Beyoncé. But the city is also increasingly multicultural. More than 20 percent of today's Houstonians were born abroad, with particularly large populations hailing from India, Vietnam, China (the city has its own flourishing Chinatown), Africa, and Latin America.

Houston also has the distinction of being the largest city in the country without zoning regulations, which its voters have repeatedly rejected, giving the city a reputation for laissez faire land management. (That said, Houston does have ordinance codes creating some restrictions.) Houston's lack of zoning has led to many businesses and houses coexisting as neighbors, creating unusual juxtapositions—and relatively affordable home prices, even as the city's population has nearly doubled since 1970, swelling to 2.3 million. As writer Nolan Gray has noted, unzoned Houston is "able to grow, adapt, and evolve like no other city" with an "ongoing supernova of construction." Perhaps America's most affordable metropolis, Houston is a car-centric city stretched over a vast flat landscape. Houston is also a major cultural and culinary destination known for its numerous museums and restaurants, as well as its large zoo and, of course, the space center—the area's top attraction for international tourists.

Reports from European explorers suggest that native tribes such as the Akokisa people once lived in what is now the Houston area. The site was sparsely inhabited in 1826, when the settler John Richardson Harris (1790–1829) founded a town within the bounds of what is now Houston and named it Harrisburg after himself. A decade later, Harrisburg was destroyed during the Texas Revolution by Mexican troops pursuing the Texas army. A week later, the Battle of San Jacinto (1836) took place about 20 miles east of present-day Houston, ending the war, and Texas gained its independence from Mexico.

The people of the newly independent Republic of Texas (1836–1846) built a town with access to the Galveston Bay navigation system to serve as a transportation hub and temporary capital. Two enterprising brothers from New York state, the investor John Kirby Allen (1810–1838) and mathematics professor-turned-businessman Augustus Chapman Allen (1806–1864), who together had worked to keep supply channels operating during the war, bought land on the banks of Buffalo Bayou for the new town. The brothers thus became the founding fathers of Houston.

The site took its name from the Virginia-born military leader, statesman, and Cherokee citizen (by induction, not birth) Sam Houston (1793–1863), who led the Texas army to victory against Mexico and was heralded as a war hero. His accomplishments over the course of his life included serving as president of the Republic of Texas, representing Texas in the U.S. Senate, becoming governor of Tennessee (although he resigned early to live among the Cherokee), and governor of Texas. He remains the only individual to ever serve as the governor of two different states.

The city of Houston served as the congressional meeting place for the Republic of Texas (1836–1846) from 1837 to 1839, when the capital moved to Austin. In 1846, Texas was formally admitted to the Union as a state, and by 1850, the first census year after Texas joined the Union, there were 2,396 Houstonians. Two decades later, that figure had grown to 9,332, and the U.S. Congress designated Houston as an official ship port. Improvements to the ship channels helped Houston thrive as a trade hub.

In 1900, disaster struck the nearby town of Galveston. A Category 4 hurricane killed between 8,000 and 12,000 Galvestonians, making it the deadliest hurricane in U.S. history to this day. Many fled inland from the ruins of the devastated Galveston, moving to Houston. The following year, oil was discovered at Spindletop, some 80 miles east of Houston. More oil was discovered in Humble, about 20 miles northeast of Houston, in 1905 and in Goose Creek, about 25 miles east of Houston, in 1906. Houston's location made it a natural choice to develop oilfield equipment.

Between the influx of new residents after the hurricane and the city's proximity to several oil site discoveries, Houston's economy grew rapidly. In 1912, Rice University was founded. In 1925, the 25-foot-deep Houston Ship Channel was completed, and Houston's port welcomed its first deep-water vessel, making the city a gateway of global trade. Houston became Texas's most populous city by the 1930 census, with 292,352 residents. The city's efficient maritime shipping made Houston rich as the Texas oil industry grew in the 1920s and 1930s, with more and more oil refineries popping up along the Houston Ship Channel. The city also developed a thriving natural gas industry. By the late 1940s, Houston's port was the second busiest in the country, ranked by tonnage of goods transported, and by the mid-1950s, Houston's population had swelled to 1 million residents.

But the 1960s are when Houston's greatest contributions to humanity arguably began, as the city became the site where flight controllers on Earth would direct astronauts to the "final frontier." After American engineer Robert Goddard (1882–1945) invented high-flying liquid-fueled rockets and American physicist J. Robert Oppenheimer (1904–1967) oversaw the first atomic bomb detonation in 1945, competition in the arena of rocketry between the Soviet Union and United States soon extended into spaceflight. The Soviet launch of the first satellite, *Sputnik 1*, in 1957 prompted the United States to create NASA. In 1961, the Soviet lead in the space race grew more pronounced when the USSR launched the first man into space, Yuri Gagarin (1934–1968), in a spacecraft called *Vostok 1*.

That year, after a lengthy search, NASA selected Houston as the location for a new manned spaceflight laboratory because of the city's warm climate, land availability, water supply, easy access to a major port, well-established industrial production capacity, and the presence of a large research university (Rice University), among other factors. The fact that the vice president at the time, Lyndon B. Johnson (1908–1973), was a Texan, may have also helped. Construction broke ground in 1962. In an address that year at Rice University, President John F. Kennedy stated:

> We meet at a college noted for knowledge, in a city noted for progress, in a State noted for strength, and we stand in need of all three. . . . This city of Houston, this State of Texas, this country of the United States was not built by those who waited and rested and wished to look behind them. This country was conquered by those who moved forward—and so will space. . . . What was once

the furthest outpost on the old frontier of the West will be the furthest outpost on the new frontier of science and space. Houston, your City of Houston, with its Manned Spacecraft Center, will become the heart of a large scientific and engineering community.

The Manned Spacecraft Center formally opened in 1963 and was renamed the Lyndon B. Johnson Space Center in 1973 after Johnson's death. The center's famed Mission Control Center has guided every American human space mission since *Gemini 4* in 1965 and manages the U.S. portions of the International Space Station today. When speaking remotely with the "CAPCOM" (the member of the operations team on the ground in charge of communications) in the Mission Control Center, astronauts refer to it by its radio call signs "Mission Control" or, simply, "Houston."

Although *Gemini 4* launched from Florida like most NASA missions, Houston assumed flight control the moment the spacecraft left the launch tower and entered the sky. The flight controllers in Houston monitored every aspect of the mission, including the spacecraft's trajectory and fuel and oxygen levels, as well as the crew's heart and breathing rates. Leading them all was the flight director—the "orchestra leader," as one retired flight controller, Sy Liebergot, put it. NASA refers to Houston as the "nerve center for American human spaceflight." *Gemini 4* was NASA's second manned Gemini spaceflight mission; it sent astronauts to orbit the Earth at a high altitude. It involved the first spacewalk (astronaut activity outside a spacecraft) by an American—less than a year after the Soviets achieved the first ever spacewalk—and conducted many science experiments.

In 1967, Houston officially adopted the Space City moniker. Flight controllers in Houston guided groundbreaking missions, including *Gemini 8* in 1966, which saw the first successful spacecraft docking, and *Apollo 8* in 1968, the first crewed mission to reach the moon and orbit it before returning to Earth. The latter's astronauts became the first people to view the entirety of the Earth from afar, a sight captured in the remarkable *Earthrise* photograph. The crew also issued a captivating Christmas Eve broadcast, reading from the book of Genesis. More people tuned in to listen to the astronauts' voices than had ever simultaneously heard any voice in history. But Houston's crowning achievement during the Space Age was undoubtedly the *Apollo 11* mission in 1969, when human feet first stepped on the moon.

The words of astronaut Neil Armstrong (1930–2012) upon touching down on the lunar surface are now famous: "That's one small step for man, one giant leap for mankind." But they were directly followed by a line directed at Houston's eponymous Mission Control, "Houston, Tranquility Base here. The *Eagle* has landed." About 600 million people, one-fifth of the global population at the time, watched the landing live, including north of 85 percent of U.S. households. People alive at the time often vividly recall where they were during that "giant leap for mankind."

The eyes of the world were on the astronauts planting an American flag on the moon. But in a windowless room, stationed in rows behind console screens relaying critical data, mostly wearing white-collared shirts with skinny ties and pocket protectors, the flight controllers in Houston were the quiet heroes of the Space Age. Their pale gray IBM consoles provided some 1,500 items of ever-changing information for analysis. Because the presence of flight controllers was required around the clock during multiday missions, each role was fulfilled in four overlapping eight-hour shifts by multiple people. At the time of the first moon landing, the average age of the flight controllers in Houston was only 32, with most having studied engineering, mathematics, or physics. The main flight director was Cliff Charlesworth (1931–1991), who held a bachelor's degree in physics and was in his late 30s.

NASA's Houston facility houses more than just Mission Control; it also once contained the Lunar Receiving Laboratory, where the first men to walk on the moon spent time quarantining upon their return to Earth, and most lunar rock samples are stored in Houston to this day. Houston also serves as a base for astronaut training.

Although crew safety always took precedence over mission success, space exploration and astronaut training are dangerous endeavors, and astronaut deaths occurred, such as that of Theodore Freeman (1930–1964), who died during astronaut training in Houston due to a bird strike. Houston takes part in NASA's annual commemoration of fallen astronauts. In a contrast of values, the Soviet government infamously concealed many space program deaths for decades, such as that of Mitrofan Nedelin (1902–1960), who perished in a covered-up launch-pad explosion along with more than 100 other people, and the Ukrainian pilot Valentin Bondarenko (1937–1961), who died during cosmonaut training at age 24.

In 1970, Houston's management skills were put to the test like never before when *Apollo 13*, the third attempted moon landing mission, suffered an oxygen tank explosion. Soon after, astronaut Jim Lovell (b. 1928) spoke the now-famous line, "Houston, we've had a problem here" (better known in the shortened form, "Houston, we have a problem" from the 1995 film *Apollo 13* that dramatized the incident).

The explosion damaged the spacecraft and made a moon landing impossible. Houston's attention turned to getting the astronauts back to Earth alive. With the command module's life support system failing, the crew moved to the lunar module. That module was only intended to support two men for two days, but thanks to innovative thinking from the team in Houston, new procedures allowed the lunar module to support three men over the course of four days. The flight director, Gene Kranz (b. 1933), chose a return route to Earth that involved looping around the moon, and Houston's Manned Spaceflight Center director Robert Gilruth (1913–2000) made decisions regarding the last part of the return journey that resulted in the safe landing of the astronauts in the Pacific. The actions of both the astronauts and the ground crew in Houston were essential to averting loss of life.

All in all, NASA completed six successful missions landing humans on the moon, with the last being *Apollo 17* in 1972. Twelve humans have walked on the moon, and all have been American astronauts whose missions were guided by Houston. The last words spoken on the moon came from astronaut Gene Cernan (1934–2017): "We leave as we came and, God willing, as we shall return, with peace, and hope for all mankind."

Despite that rhetoric emphasizing unity, international rivalry was a major factor motivating space exploration. After the

Cold War's end, the space industry was no longer subject to the intense competition that drove progress during the space race, and crewed space exploration stagnated. As of this writing, only four moonwalkers remain alive, ranging in age from 87 to 93. But a new era of private space endeavors—led by companies such as SpaceX, Blue Origin, and Virgin Galactic—may once again allow humanity to reach for the stars as profits drive a new space race. Nearly 400 miles southwest of Houston, near the southernmost tip of Texas, SpaceX has constructed its own spaceport, Starbase. Today, Houston also houses an urban commercial spaceport that is expanding as Space City seeks to position itself as a hub not just for NASA activity but also for private spaceflight.

Houston grew from a war-ravaged, struggling trading post into a global oil-shipping nexus and then the capital of the Space Age. From liftoff onward, the American astronauts who shattered records and tested the limits of the possible relied on Houston to ensure mission success and bring them safely home. Many people still consider the moon landing to be among the greatest achievements of humanity. It was certainly the farthest-reaching feat of exploration in history. For guiding mankind to the final frontier, Houston has landed in these pages as a center of progress.

38

Berlin

FALL OF COMMUNISM

Berlin played a central role in the fall of communism and the triumph of liberalism. When the wall that had divided Berlin was abruptly and joyfully torn down in 1989, the city changed human history.

Today, Berlin is the most populous city in the European Union, with about 3.8 million residents. Famed for its history, art, music, and graffiti, Berlin attracts millions of tourists each year, as well as many business travelers. The city's economy revolves around the high-tech and service industries, and the metropolis is a major transportation hub.

The site where Berlin now stands has been inhabited since at least the ninth millennium BCE with many artifacts such as arrowheads surviving from ancient villages in the area. During the Bronze Age and Iron Age, the primary residents were members of the Lusatian culture, an agricultural people notable for cremating rather than burying their dead. Various tribes migrated through the region, and by the seventh century, Slavic people populated the area. Berlin's name likely means "swamp" in Polabian, a now-extinct Slavic language.

The similarity between the city's name and the modern word "bear" (*bär* in German), along with the bear on the city's coat of arms, has led to a popular misconception that the city is named after the animal. The coat of arms was actually given to the city by a nobleman known as Albert the Bear, who took control of the area in the 12th century when he established the Margraviate of Brandenburg in 1157.

Officially founded in 1237 (although in fact established before that), Berlin endured a tumultuous couple of centuries. Despite a devastating fire in 1380, Berlin managed to reach a population of about 4,000 residents by 1400. Berlin then suffered considerable damage during the Thirty Years' War (1618–1648) but again rebounded, seeing a burst of growth after becoming the capital of the new Kingdom of Prussia in the 18th century. As the seat of Prussian power, the city was a center of administration and entrepreneurship. Workshops sprang up, and Berlin became known for its skilled craftsmen.

By the 19th century, limited access to power generated by waterwheels forced the city to adopt steam power early. Harnessing steam energy allowed Berlin to industrialize rapidly and become a major producer of everything from clothing and chemicals to heavy machinery. The city's central location made it Germany's rail transportation hub, and Berlin was soon an economic powerhouse.

As the city grew prosperous, it became a sanctuary for the German Romantic movement, hosting painters, musicians, poets, and writers. The Austrian-born Romantic composer Franz von Suppé (1819–1895) is alleged to have written lyrics that translate to "You are crazy my child, you must go to Berlin / where the crazy ones are / there you belong." While those lyrics (made famous by citation in a 1958 film, *Der eiserne Gustav*) are likely a later addition to a melody that Suppé composed, they nonetheless capture the creative spirit that took hold of the city. Berlin soon gained a reputation as a home for artistic misfits from across the continent.

In the 20th century, Berlin maintained that reputation as German expressionist painters and filmmakers experimented with new styles in the city. Despite growing economic and political instability throughout the Weimar Republic, Berlin was a renowned nightlife and creative center during the Roaring Twenties. The city's thinkers also made notable contributions to science, and its universities gained increasing prominence. The physicist Albert Einstein (1879–1955) won the Nobel Prize in Physics in 1921 while working at Berlin's Humboldt University.

The intellectual freedom that pervaded the city was suddenly and dramatically extinguished with the rise of National Socialism (Nazism) and the establishment of the totalitarian Third Reich (1933–1945). Many of the artists and scientists who had put the city on the map, including Einstein, fled Berlin to escape the genocidal rule of Adolf Hitler (1889–1945). After Hitler's defeat at the end of World War II, the Allies divided Germany into four different occupation zones. The Soviet Union gained control of the eastern part of Berlin and declared the city the capital of the new Soviet satellite state in East Germany.

East Germany's official name was the German Democratic Republic. Its government was modeled after the Soviet Union's, complete with central planning, state ownership of the means of production, limits on private property, de facto single-party rule, censorship, a vast spy and repression network, and an ostensible commitment to class equality.

West Berlin and West Germany quickly recovered from World War II and grew wealthy, but the tight government controls on East Germany's economy prevented a similar recovery. Although perhaps history's best natural experiment testing capitalism against communism, the partition was devastating for the people of East Germany. Between 2.5 and 3.0 million East Germans escaped to the West. By 1961, it is thought that about 1,000 East Germans fled daily, many through Berlin. Those with advanced education or professional skills were particularly likely to make a run for freedom. As the young socialist state hemorrhaged many of its brightest citizens, its leaders grew desperate. Walter Ulbricht, the chief decisionmaker in East Germany, received the blessing of the Soviet premier Nikita Khrushchev to stop the outflow with a physical barrier.

In August 1961, soldiers erected a barbed-wire barricade to block access from East Berlin to West Berlin. The wire barrier was then replaced by an enormous wall. The Berlin Wall was made of solid concrete blocks, stood 6 feet tall, and ran for

96 miles. Officers known as Volkspolizei ("Volpos") manned the wall's guard towers, searchlights, and machine-gun posts at all times. The barrier separated families and friends.

A secret police force called the Stasi, headquartered in East Berlin, monitored citizens' private lives to detect and prevent escape plans or any activity that might challenge communist rule. The Stasi's mass surveillance campaign included covertly reading all mail sent through the state-run postal system, setting up a vast network of informants and installing wiretaps in the homes of numerous citizens.

The Stasi sought to psychologically destroy dissidents identified by its spies through a program known as *Zersetzung* (decomposition). Stasi operatives manipulated victims' lives to disrupt their careers and all of their meaningful personal relationships (such as by planting false evidence of adultery into a couple's life). The goal was for the victim to wind up companionless, a social and professional failure, and utterly lacking in self-esteem. The program is thought to have involved up to 10,000 victims and irreversibly damaged at least half of those victims' minds. (Today, recognized *Zersetzung* survivors receive special pensions.)

Despite the risks, the frequent material shortages and relative poverty generated by the dysfunctional communist system motivated a continuous stream of East Germans to attempt escape. Between 1961 and 1988, well over 100,000 East Germans tried to cross the Berlin Wall, but almost all were apprehended and imprisoned. At least 600 were gunned down or otherwise killed during the attempt to flee to the West. Only about 5,000 crossed successfully in the 27-year period.

On June 26, 1963, U.S. President John F. Kennedy delivered what is considered one of history's greatest speeches in West Berlin. His words resonated with Berliners:

> There are many people in the world who really don't understand, or say they don't, what is the great issue between the free world and the communist world. Let them come to Berlin! There are some who say that communism is the wave of the future. Let them come to Berlin! . . . Freedom has many difficulties and democracy is not perfect, but we have never had to put a wall up to keep our people in, to prevent them from leaving us . . . [T]he wall is the most obvious and vivid demonstration of the failures of the communist system . . . . All free men, wherever they may live, are citizens of Berlin and, therefore, as a free man, I take pride in the words *"Ich bin ein Berliner!"*

Although East Berliners dreamed of escape, West Berlin thrived and once again attracted groundbreaking artists and musicians. In the late 1970s, the English singer David Bowie called West Berlin "the greatest cultural extravaganza that one could imagine." His 1977 song "Heroes," written in Berlin and inspired by the sight of a couple embracing by the Berlin Wall, has since become an unofficial anthem of the city and of resistance to totalitarianism more broadly. (After the singer's death in 2016, the German government even recognized the song's impact and thanked Bowie for his role in "helping to bring down the Wall.") Other musical successes of West Berlin during the period include the 1983 antiwar anthem "99 Luftballons."

Opposition to the Berlin Wall continued to mount. In 1987, during a speech in West Berlin, the then U.S. president

Ronald Reagan famously called on the Soviet leader to remove the barrier, saying, "Mr. Gorbachev, tear down this wall!"

On November 9, 1989, as socialism's unviability became increasingly hard to deny and the Cold War thawed, East Berlin's Communist Party spokesperson unexpectedly announced that crossing the Berlin Wall would be legal at midnight. A tidal wave of East and West Berliners rushed to the wall, chanting "*Tor auf*!" (Open the gate!). At midnight, long-separated friends, family members, and neighbors flooded across the barrier to reunite and celebrate.

More than 2 million East Berliners are believed to have crossed into West Berlin that weekend, resulting in what one journalist described as "the greatest street party in the history of the world." Revelers joyously graffitied and smashed apart the wall with hammers while bulldozers demolished other sections.

The fall of the Berlin Wall symbolized the end of widespread support for communism and a global turn toward policies of greater economic and political freedom. "For West Germans, nothing changed other than postcodes. For East Germans, everything changed," as one German living in the former East told Reuters.

The city was reunited, but even today, the economic and psychological scars of the Cold War partition can be felt. East Berlin is still plagued by higher levels of dishonesty (according to research by Duke University behavioral scientist Dan Ariely and several coauthors in the *European Journal of Political Economy*) and lower levels of trust than West Berlin, although East Berliners have mostly caught up to their West Berlin counterparts when it comes to life satisfaction.

The story of Berlin reads like a parable about the importance of freedom. The breaching of the wall not only freed millions of Germans from poverty and despotism but proved to be a pivotal moment in history that helped millions of other people achieve greater economic and political freedom as well. For tearing down the wall, Berlin has won its place as a center of progress.

39

Tokyo

TECHNOLOGY

Our next center of progress is Tokyo, which, after it was nearly destroyed during World War II, was rapidly rebuilt and reinvented itself as a world leader in manufacturing and technology.

Today, Tokyo is its country's economic center and the seat of the Japanese government. The city is famously safe and prosperous. It is renowned for its glamour and cosmopolitanism. The greater Tokyo area is currently the most populous metropolitan area in the world, boasting well over 37 million residents. As our penultimate center of progress, Tokyo's large population is appropriate because, as in every city, it is the people who live or have lived there who have driven progress and created wealth. And the more people, the merrier—a finding also backed up by empirical research, such as the findings by Marian L. Tupy and Gale Pooley that higher populations, counterintuitively, can make resources more abundant.

Situated on Tokyo Bay, the metropolis began as a humble fishing village. Originally named Edo, meaning "estuary," the area first rose to prominence when it was designated as the seat of the Tokugawa shogunate in 1603. By the 18th century, the once-obscure locale had grown into one of the world's most populous cities with a population of over 1 million people.

The city benefited from a lengthy peace known as the *Pax Tokugawa*, which let the city's people devote their resources to economic development rather than military defense. That was particularly fortunate because the city often had to be rebuilt after disasters. The city was vulnerable to fires, thanks to its predominantly wooden architecture, as well as earthquakes—a consequence of Japan's location along the so-called Ring of Fire, the most earthquake-prone zone on Earth. Tokyo's ability to thrive when spared from the vicissitudes of conflict is a recurring theme in the city's story.

When the Tokugawa shogunate ended in 1868, the newly empowered imperial court moved to Edo and renamed the city Tokyo, meaning "eastern capital," a reference to the previous capital city of Kyoto, which is located nearly 300 miles to Tokyo's west. As the headquarters of the new regime, Tokyo was at the forefront of the Meiji Restoration (1868–1912), an era of Japanese history characterized by rapid modernization. In just a few decades, the country abolished feudal privileges and industrialized its economy, becoming a modern state complete with paved roads, telephones, and steam power. During the subsequent Taishō era (1912–1926), Tokyo continued to expand as Japan urbanized and modernized further.

In 1923, disaster struck the city. The Great Kantō earthquake, which measured 7.9 on the Richter scale, caused a fire whirl that burned down the city center. More than 140,000 people perished in the catastrophe, and about 300,000 homes were destroyed. At the time, it was the worst tragedy the city had ever experienced. But just over two decades later, the catastrophe was superseded by the far worse devastation wrought by World War II.

Japan was among the countries most devastated by World War II, losing between 1.8 and 2.8 million people as well as a quarter of the nation's wealth. The country sustained damage not only from the nuclear bombs dropped on Hiroshima and Nagasaki but also from an extremely effective campaign of conventional bombing of some of its biggest cities, including Nagoya, Osaka, Kobe—and Tokyo. Operation Meetinghouse (March 1945), or the Great Tokyo Air Raid, is considered to be the single most destructive bombing raid of World War II. It was deadlier than the bombings of Dresden or Hamburg and even the nuclear attacks on Hiroshima or Nagasaki.

The low-altitude incendiary raid claimed the lives of at least 100,000 Tokyoites, wounded over 40,000 others, burned a quarter of the city to the ground, and left 1 million people homeless. Temperatures reached 1,800 degrees Fahrenheit on the ground in some parts of Tokyo, and the city's mainly wooden structures quickly disappeared in flames. And that was just one of the multiple firebombings the city suffered during the war. In addition to being the target of World War II's deadliest bombing, Tokyo was also a target of what was likely the single-largest bombing raid in history, involving more than 1,000 planes.

The firebombings collectively cut Tokyo's economic output in half. As a whole, Japan's industrial production was reduced to a tenth of its prewar levels. Industrial and commercial buildings and machinery were particularly likely to have been destroyed during the war.

That destruction contributed to wide-ranging postwar food and energy shortages, and infrastructure damage made transportation to some areas almost impossible. Combined with the abrupt demobilization of the country's 7.6 million soldiers, approximately 4 million civilians engaged in war-related work, and 1.5 million returnees from territories that Japan occupied during the war, the devastation contributed to already massive unemployment. With over 13 million people out of work in the country as a whole, rampant inflation, and currency devaluation, Tokyo's economy came to an effective standstill.

Despite the grim situation, postwar Tokyo also had a few advantages that favored quick recovery. Prewar Japan had been a major power. The capital city maintained an institutional memory of what it was like to be an industrial center and still possessed an educated and skilled workforce. The American Occupation Administration was also highly motivated to help with the economic turnaround, as the United States was invested in seeing the country's swift demilitarization and democratization.

The United States forced Japan to surrender its right to a military and assumed the cost of the country's defense, thus allowing Japan to allocate its full resources toward civilian activities, such as commercial investment. Many Japanese leaders, such as Prime Minister Shigeru Yoshida (1878–1967), fully supported demilitarization. He is sometimes called the father of the modern Japanese economy. Even after Japan established a national defense force in 1954, the expense was small and shrank as a portion of gross domestic product over the years. Some economists estimate that Japan's economy would have been 30 percent smaller by 1976 if it had not been freed from the burden of military spending. (Conversely, higher U.S. military spending from funding Japan's defense may have slowed economic growth in the United States to some degree.)

Japan swiftly enacted several economic reforms. The Allies compelled the country to disband the *zaibatsu*, the crony-capitalist conglomerates that had received preferential treatment from the imperial government, ranging from lower tax rates to cash stimulus injections. Because of their entanglement with the government, the zaibatsu had managed to maintain a near-monopoly over vast swaths of the economy and crush competitors. Ending the reign of the zaibatsu allowed new companies to form and compete in a more open economy. At the same time, Japan passed land reforms that transformed the country's agriculture, which had been previously operating along inefficient feudal lines.

As the Cold War began in the late 1940s, the United States hoped that Japan would become a strong democratic and capitalist ally in the region. To that end, in 1949, the banker and U.S. presidential adviser Joseph Dodge (1890–1964) helped Japan balance its budget, bring inflation under control, and remove widespread government subsidies that propped up inefficient practices. Dodge's policies, now known as the Dodge Line, decreased the level of state intervention in the Japanese economy, making the latter much more dynamic. Shortly after those policies took effect, the Korean War (1950–1953) broke out, and the United States procured many of its war supplies from the geographically close Japan. Economic liberalization, combined with the sudden increase in manufacturing demand, supercharged Japan's, and particularly Tokyo's, recovery.

Tokyo began to experience mind-bogglingly fast economic growth. The city rapidly reindustrialized and acted as a major trade hub as the country's imports and exports increased dramatically. The archipelago nation had relatively few natural resources, but by importing large amounts of raw materials to manufacture finished goods, Japan was able to achieve impressive economies of scale, multiply manufacturing output, and increase profits. Those profits were then reinvested into better equipment and technological research, boosting output and profits in a virtuous cycle.

The U.S. government removed trade barriers on Japanese goods and, by and large, resisted calls to institute protectionist measures against Japan and encouraged other nations to do the same, thus ensuring that Japanese entrepreneurs were free to sell their goods in the United States and elsewhere. In the post–Korean War period, banks in the United States and elsewhere invested heavily in Japan's economy and expected large returns.

They were rewarded as Japan's economic miracle materialized and Tokyo flourished. Between 1958 and 1960, Japanese exports to the United States increased by 150 percent. In 1968, less than 22 years after World War II, Japan boasted the world's second-largest economy, and Tokyo was at the heart of the nation's newfound prosperity.

Tokyo soon became the birthplace and home base of major global companies, producing cars (Honda, Toyota, Nissan, Subaru, and Mitsubishi), cameras (Canon, Nikon, and Fujifilm), watches (Casio, Citizen, and Seiko), and other digital goods (Panasonic, Nintendo, Toshiba, Sony, and Yamaha).

Tokyo's entrepreneurial success is partly due to innovation. Toyota, for example, edged out American car manufacturers by creating a new production system that used strategic automation and "just-in-time manufacturing," thus increasing efficiency. Just-in-time manufacturing—a production model where each step in the manufacturing process is timed to eliminate the need for excess inventory storage—has since become the global norm across a range of industries.

Since the 1970s, Tokyo has also become renowned for cutting-edge robotics. Developing expertise in industrial robotics was a natural extension of the city's manufacturing prowess, but Tokyo companies and researchers have since branched out into many other areas of robotics. The city has created innovations ranging from robotic bellhops and airport greeters to friendly robotic baby seals that assist Alzheimer's patients.

Sadly, since the early 1990s, due to a variety of misguided government policies, Japan's economy has seen a significant slowdown and suffered from decades-long stagnation. The country's bygone era of rapid economic expansion is nonetheless worth studying.

The largely destroyed capital city of a country devastated by war managed to transform itself into one of the world's leading technology centers within a few decades. Thanks to the ingenuity and determination of the city's people combined with conditions of peace, economic freedom, and the opportunity to engage in global trade, Tokyo became an economic miracle that qualifies it as one of modern history's great urban success stories. And it is fitting that a city at the forefront of technological progress should be counted as a center of progress.

40

San Francisco

DIGITAL REVOLUTION

Our last center of progress is San Francisco during the digital revolution, when entrepreneurs founded several major technology companies in the area. The southern portion of the broader San Francisco Bay Area earned the metonym Silicon Valley because of the high-tech hub's association with the silicon transistor, used in all modern microprocessors. A microprocessor is the central unit or engine of a computer system, fabricated on a single chip.

Humanity has long strived to develop tools to improve our survival odds and make our lives easier, more productive, and more enjoyable. In the long history of inventions that have made a difference in the average person's daily life, digital technology, with its innumerable applications, stands out as one of the most significant innovations of the modern age.

Today, the Bay Area remains best known for its association with the technology sector. With its iconic Victorian houses, sharply sloping hills, streetcars, fog, Chinatown (which bills itself as the oldest and largest one outside Asia), and, of course, the Golden Gate Bridge, the city of San Francisco is famous for its distinctive views. As the *Encyclopedia Britannica* notes, "San Francisco holds a secure place in the United States' romantic dream of itself—a cool, elegant, handsome, worldly seaport whose steep streets offer breathtaking views of one of the world's greatest bays." Attempts to preserve the city's appearance have contributed to tight restrictions on new construction. Perhaps relatedly, the city is one of the most expensive in the United States and suffers from a housing affordability crisis. San Francisco has in recent years struggled with widespread homelessness and related drug overdose deaths and crime. With both the country's highest concentration of billionaires—thanks to the digital technology industry—and the ubiquitous presence of unhoused people, San Francisco is a city of extremes.

Today's densely populated metropolis was once a landscape of sand dunes. In 1769, the first documented sighting of the San Francisco Bay was recorded by a scouting party led by the Spanish explorer Gaspar de Portolá (1716–1786). In 1776, European settlement of the area began, led by the Spanish missionary Francisco Palóu (1723–1789) and expeditionary José Joaquín Moraga (1745–1785). The latter is the namesake of San Jose, a city on the southern shore of San Francisco Bay, about 50 miles from San Francisco but located within the Bay Area and the San Jose–San Francisco–Oakland combined statistical area. San Francisco was the northernmost outpost of the Spanish Empire in North America and later the northernmost settlement in Mexico after that country's independence. But the city remained relatively small and unknown.

In 1846, during the Mexican-American War, the United States captured the San Francisco area, although Mexico did not formally cede California until the 1848 Treaty of Guadalupe Hidalgo. At that time, San Francisco had only about 900 residents. That number grew rapidly during the California Gold Rush (1848–1855), when the discovery of gold turned the quiet village into a bustling boomtown of tens of thousands by the end of the period. Development of the city's port led to further growth and helped the area become a hub in the early radio and telegraph industries, foreshadowing the city's role as a leader in technology.

In 1906, three-quarters of the city was destroyed in a devastating earthquake and related fire caused by a gas line rupturing in the quake. The city rebuilt from the destruction and continued its growth, along with the broader Bay Area. In 1909, San Jose became the home of one of the first radio stations in the country. In the 1930s, the Golden Gate Bridge became a part of San Francisco's skyline, and the city's storied Alcatraz maximum security prison opened, holding famous prisoners such as the Prohibition-era gangster Al Capone (1899–1947). In 1939, in Palo Alto, just over 30 miles south of San Francisco, William Hewlett (1913–2001) and David Packard (1912–1996) founded a company that made oscilloscopes—laboratory instruments that display electronic signals as waves. They named the company Hewlett-Packard. During World War II, the company shifted to making radar and artillery technology. That field soon became linked to computing. That link occurred because researchers at the University of Pennsylvania created a new tool to calculate artillery firing tables, among other tasks: the first general-use digital computer.

"Computer" was once a job title for a person who performed calculations. The first machine computer, named ENIAC (Electronic Numerical Integrator and Computer), debuted in 1945. It cost about $500,000, or nearly $8 million in 2022 dollars, measured 8 feet tall and 80 feet long, weighed 30 tons, and needed constant maintenance to replace its fragile vacuum tubes. Back when computers were the size of a room and required many people to operate them, they also had about 13 times less power than a modern pocket-sized smartphone that costs about 17,000 times less.

San Francisco and Silicon Valley's greatest claim to fame came with the dawn of more convenient and powerful digital technology. In 1956, the inventor William Shockley (1910–1989) moved from the East Coast to Mountain View, a city on San Francisco Bay located about 40 miles south of San Francisco, to live closer to his ailing mother. She still lived in his childhood home of Palo Alto. That year he won the Nobel Prize in Physics along with engineer John Bardeen (1908–1991) and physicist Walter Houser Brattain (1902–1987). The prize honored them for coinventing the first working semiconductor almost a decade earlier, in 1947, at Bell Laboratories in New Jersey.

After moving to California, Shockley founded Shockley Semiconductor Laboratory, the first company to make transistors and computer processors out of silicon—earlier versions used germanium, which cannot handle high temperatures. His work provided the basis for many further electronic developments.

Also in 1956, IBM's San Jose labs invented the hard-disk drive. That same year, Harry Huskey (1916–2017), a professor at the University of California, Berkeley, some 14 miles from San Francisco, designed Bendix's first digital computer, the G-15.

Shockley had an abrasive personality and later became a controversial figure because of his vocal fringe views related to eugenics and mass sterilization. In 1957, eight of Shockley's employees left over disagreements with Shockley to start their own enterprise together with investor Sherman Fairchild (1896–1971). They named it Fairchild Semiconductors. Shockley called them the Traitorous Eight. In the 1960s, Fairchild Semiconductors made many of the computer components for the Apollo space program directed from Houston, one of our previous centers of progress. In 1968, two of the Traitorous Eight—Gordon Moore (b. 1929) and Robert Noyce (1927–1990), the latter of whom earned the nickname the Mayor of Silicon Valley—left Fairchild to start a new company in Santa Clara, about 50 miles southeast of San Francisco. They named it Intel. Moore remains well-known as the creator of Moore's law. It was he who predicted in 1965 that the processing power of computers would double every 18 months.

In 1969, the Stanford Research Institute at Stanford University, some 35 miles southeast of San Francisco, became one of the four "nodes" of the Advanced Research Projects Agency Network. ARPANET was a research project that would one day become the internet. In 1970, Xerox opened the PARC laboratory in Palo Alto, which would go on to invent ethernet computing and graphic user interfaces. In 1971, journalist Don Hoefler (1922–1986) published a three-part report on the burgeoning computer industry in the southern San Francisco Bay Area that popularized the term "Silicon Valley." The pace of technological change picked up with the invention of microprocessors that same year.

Just as the 19th-century Gold Rush once attracted fortune seekers, the promise of potential profit and the excitement of new possibilities offered by digital technology drew entrepreneurs and researchers to the San Francisco Bay Area. In the 1970s, companies such as Atari, Apple, and Oracle were all founded in the area. By the 1980s, the Bay Area was the undisputed capital of digital technology. (Some consider the years from 1985 to 2000 to constitute the golden era of Silicon Valley, when legendary entrepreneurs such as Steve Jobs [1955–2011] were active there.) San Francisco suffered another large earthquake in 1989, but that was accompanied by a relatively small death toll. In the 1990s, companies founded in the Bay Area included eBay, Yahoo!, PayPal, and Google. The following decade, Facebook and Tesla joined them. As these companies created value for their customers and achieved commercial success, fortunes were made, and the San Francisco Bay Area grew wealthier. That was particularly true of San Francisco.

Although many of the important events of the digital revolution took place across a range of cities in the Bay Area, San Francisco itself was also home to the founding of several significant technology companies. Between 1995 and 2015, major companies founded in or relocated to San Francisco included Airbnb, Coinbase, Craigslist, DocuSign, DoorDash, Dropbox, Eventbrite, Fitbit, Flickr, GitHub, Grammarly, Instacart, Instagram, Lyft, Niantic, OpenTable, Pinterest, Reddit, Salesforce, Slack, TaskRabbit, Twitter, Uber, WordPress, and Yelp.

San Francisco helped create social media and the so-called sharing economy that offers many workers increased flexibility. By streamlining the process of such things as grocery deliveries, restaurant reservations, vacation home rentals, ride-hailing services, secondhand sales, cryptocurrency purchases, and work group chats, enterprises based in San Francisco have made countless transactions and interactions far more convenient.

New technologies often present challenges as well as benefits, and the innovations of San Francisco along with Silicon Valley are certainly no exception. Concerns about data privacy, cyberbullying, social media addiction, and challenges related to content moderation of online speech are just some of the issues attracting debate today that relate to digital technology. But there is no going back to a world without computers, and most would agree that the immense gains from digital technology outweigh the various dilemmas posed by it.

San Francisco's golden age has ended. As the city struggles to address its homelessness crisis and various other problems, numerous technology companies have relocated and many technological breakthroughs are now occurring elsewhere. Do not take San Francisco's placement at the end of this book to mean that the city today presents a model to be emulated. Quite the opposite. Even so, the area's erstwhile achievements merit celebration.

Practically everyone with access to a computer or smartphone has direct experience benefiting from the products of several San Francisco companies, and the broader San Francisco Bay Area played a role in the very creation of the internet and modern computers. It is difficult to summarize all the ways that computers, tablets, and smartphones have forever changed how humanity works, communicates, learns, seeks entertainment, and more. There is little doubt that San Francisco has been one of the most innovative and enterprising cities on Earth, helping define the rise of the digital age that transformed the world. For these reasons, San Francisco is this book's last center of progress . . . but not humanity's.

There are doubtless many more world-changing innovations to come. Thomas Jefferson once quipped, "I like the dreams of the future better than the history of the past." What will be the next great center of progress? No one can say for certain. You might be standing in it.

Acknowledgments

I began writing the online series of articles that became this book in April 2020, not long after the COVID-19 pandemic sent the Washington, DC, area, where I live, into a soft lockdown. Learning about each featured city was a nice escape while stuck at home, social-distancing. This book would not have been possible without the support of my husband, Drew, who brought me coffee and gave me time to write by watching our feisty toddler. My life changed considerably over the course of writing. My family moved to a new home and welcomed our second child. This project provided a constant through a period of dramatic transitions.

I thank Marian Tupy for his wisdom, guidance, good humor, and many years of professional mentorship, and Ian Vásquez for his belief in the project from its inception. I also thank my colleagues Aaron Steelman, Eleanor O'Connor, and Ivan Osorio for shepherding this project to fruition and my former colleague Jason Kuznicki for his helpful comments. I am grateful to Malcolm Cochran and David Behrens for their help in editing the initial drafts, as well as Cato's editorial team, particularly senior copy editor Karen Garvin. I thank my colleagues Saul Zimet and Walker Haskins for their assistance in fact-checking the volume, and my fellow Cato scholars Mustafa Akyol, Terence Kealey, and José Piñera for their words of encouragement and for sharing their insights on several of the cities.

Thank you to Yuriy Romanovich, whose illustrations for the initial web series of articles have translated beautifully to the pages of this book, and who worked from his native Ukraine even as the country came under attack.

I am grateful to Luis Ahumada Abrigo, who promoted the series; to high school teacher Sean Kinnard for crafting lesson plans based on it; and to the Sphere Education Initiative's Allan Carey and Elyse Alter for helping get those lesson plans—and this book—into classrooms. *Gracias* to Gabriela Calderon de Burgos for translating the city profiles into Spanish for ElCato.org. I thank my former colleague Alexander C. R. Hammond, whose "Heroes of Progress" profiles helped inspire this book and will form its companion volume. And I thank readers Gerald O'Driscoll and Michael Stay for reaching out and suggesting the inclusion of Bologna and Göbekli Tepe, respectively, as well as the many other readers who offered kind words and city suggestions that I wasn't able to fit into the final work. I also owe gratitude to Luis (again) and Guillermina Sutter Schneider for the book's design.

I thank my father and late mother for their faith in me, and my children for the joy and sense of meaning they lend to my life. And most of all, I thank my husband, who pushed me to meet my deadlines and gave me constant encouragement.

Illustrations & Figures

Chapter 4. Uruk (Writing)

Evolution of the cuneiform sign for "head" during the third millennium BCE, after Samuel Noah Kramer, *Thirty Nine Firsts In Recorded History* (Philadelphia: University of Pennsylvania Press, 1988) via Wikimedia Commons, uploaded by user Dbachmann. The image has been made monochromatic and truncated so that the last iteration of the sign represents the late third millennium BCE.

Chapter 21. Mainz (Printing Press)

Fifteenth-century printing towns of incunabula, based on the Incunabula Short Title Catalogue of the British Library via Wikimedia Commons, uploaded by user NordNordWest.

Chapter 30. Manchester (Industrialization)

Real GDP per capita, 2011 U.S. dollars. HumanProgress.org.

Chapter 33. Chicago (Railroads)

Raising of the Briggs House, a Chicago hotel, in 1857. From the Chicago Historical Society.

Chapter 36. Hong Kong (Noninterventionism)

Real average income in Hong Kong and Great Britain, 1945–1997. HumanProgress.org.

Suggested Further Reading

Overall

Bailey, Ronald, and Marian L. Tupy, *Ten Global Trends Every Smart Person Should Know: And Many Others You Will Find Interesting* (Washington: Cato Institute, 2020).

McCloskey, Deirdre, "How the West (and the Rest) Got Rich," *Wall Street Journal*, May 20, 2016.

Norberg, Johan, *Progress: Ten Reasons to Look Forward to the Future* (New York: Oneworld Publications, 2017).

Norwich, John Julius, *Cities That Shaped the Ancient World* (London: Thames & Hudson, 2014).

———, *The Great Cities in History* (London: Thames & Hudson, 2009).

Ridley, Matt, *The Evolution of Everything: How New Ideas Emerge* (New York: Harper, 2015).

Tupy, Marian L., and Gale Pooley, *Superabundance: The Story of Population Growth, Innovation, and Human Flourishing on an Infinitely Bountiful Planet* (Washington: Cato Institute, 2022).

Chapter 1. Jericho (Agriculture)

Joyce, Christopher, "Ancient Figs May Be First Cultivated Crops," NPR, June 2, 2006.

Parry, Wynne, "Mystery of Ancient Jericho Monument Revealed," CBS News, February 18, 2011.

Scott, James C., *Against the Grain: A Deep History of the Earliest States* (New Haven, CT: Yale University Press, 2017).

University of Sheffield, "Why Did Hunter-Gatherers First Begin Farming?," *ScienceDaily*, May 16, 2017.

Chapter 2. Göbekli Tepe (Religion)

Dietrich, Oliver, Manfred Heun, Jens Notroff, Klaus Schmidt, and Martin Zarnkow, "The Role of Cult and Feasting in the Emergence of Neolithic Communities: New Evidence from Göbekli Tepe, South-Eastern Turkey," *Antiquity* 86 no. 333 (2012): 674–95.

Mann, Charles C., "The Birth of Religion," *National Geographic*, June 2011, pp. 34–59.

Thomas, Sean, "Is an Unknown, Extraordinarily Ancient Civilisation Buried under Eastern Turkey?," *The Spectator*, May 8, 2022.

Chapter 3. Budj Bim (Aquaculture)

Bellware, Kim, "Ancient Aboriginal Aquaculture System Older than Stonehenge Uncovered by Australian Wildfires," *Washington Post*, January 21, 2020.

Gattuso, Reina, "Australian Wildfires Uncovered Hidden Sections of a Huge, Ancient Aquaculture System," *Atlas Obscura*, February 6, 2020.

McDougall, Rennie, "Aboriginal Aquaculture: The First Australian Engineers and How History Almost Overlooked Them," *Lapham's Quarterly*, August 6, 2018.

Chapter 4. Uruk (Writing)

Ghazal, Rym, "World's Oldest Writing Not Poetry but a Shopping Receipt," *National News UAE*, April 12, 2011.

Glassner, Jean-Jacques, *The Invention of Cuneiform: Writing in Sumer*, translated by Zainab Bahrani and Marc Van De Mieroop (Baltimore: Johns Hopkins University Press, 2003).

Harford, Tim, "How the World's First Accountants Counted on Cuneiform," BBC, June 12, 2017.

University of Oxford, "Enmerkar and the Lord of Aratta: Translation," Electronic Text Corpus of Sumerian Literature.

Chapter 5. Mohenjo-Daro (Sanitation)

Carter, W. Hodding, *Flushed: How the Plumber Saved Civilization* (New York: Atria Books, 2007).

Roach, John, "Mohenjo-Daro 101," *National Geographic*, October 9, 2009.

Robinson, Andrew, "The Real Utopia: This Ancient Civilisation Thrived without War," *New Scientist*, September 4, 2016.

Watson, Traci, "Surprising Discoveries from the Indus Civilization," *National Geographic*, April 30, 2013.

Chapter 6. Nan Madol (Seafaring)

Moulton, Madison, "What Is the Austronesian Expansion?," *History Guild*, March 12, 2021.

Pala, Christopher, "Nan Madol: The City Built on Coral Reefs," *Smithsonian Magazine*, November 3, 2009.

Thompson, Christina, *Sea People: The Puzzle of Polynesia* (New York: Harper, 2019).

Chapter 7. Memphis (Medicine)

Rossi, Marco, "Homer and Herodotus to Egyptian Medicine," *Vesalius: acta internationales historiae medicinae*, no. S3–5, 2010, PMID: 21657099.

Stiefel, Marc, Arlene Shaner, and Steven D. Schaefer, "The Edwin Smith Papyrus: The Birth of Analytical Thinking in Medicine and Otolaryngology," *Laryngoscope* 116, no. 2 (2006): 182–88.

University of Manchester, "Egyptians, Not Greeks, Were True Fathers of Medicine," May 9, 2007.

Chapter 8. Ur (Law)

Code of Ur-Nammu.

Editors of *Encyclopedia Britannica*, "Cuneiform Law," *Encyclopedia Britannica*, January 21, 2011.

Kramer, Samuel Noah, "Law and Love: A Hymn, a Prayer, and a Word to the Wise," *Penn Museum Bulletin* 18, no. 2 (1952): 23–42.

Roth, Martha T., *Law Collections from Mesopotamia and Asia Minor,* vol. 6, edited by Piotr Michalowski (Atlanta: Scholars Press, 1995), pp. 16–21.

Chapter 9. Chichén Itzá (Team Sports)

Blakemore, Erin, "Where Did Soccer Start? Archaeology Weighs In," *National Geographic*, June 15, 2018.

Taylor, Steve, "Sport and the Decline of War," *Psychology Today*, March 14, 2014.

Whittington, E. Michael, ed., *The Sport of Life and Death: The Mesoamerican Ballgame* (London: Thames & Hudson, 2001).

Chapter 10. Athens (Philosophy)

Long, Roderick T., "Athens, for All Its Flaws, Was a Beacon of Personal Liberty in the Ancient World," Libertarianism.org, September 24, 2015.

Merchant, E. C., ed., "Pseudo-Xenophon (Old Oligarch)," *Xenophon in Seven Volumes* (Cambridge, MA: Harvard University Press, 1984).

Weiner, Eric, "Genius Is Simple: Athens," *The Geography of Genius: Lessons from the World's Most Creative Places* (New York: Simon & Schuster, 2016), pp. 13–64.

Chapter 11. Alexandria (Information)

El-Abbadi, Mostafa, "Library of Alexandria," *Encyclopedia Britannica*, July 17, 2020.

Garlinghouse, Tom, "The Rise and Fall of the Great Library of Alexandria," LiveScience, March 14, 2022.

Chapter 12. Rome (Roads)

Editors of *Encyclopedia Britannica*, "Roman Road System," *Encyclopedia Britannica*, April 3, 2018.

Malacrino, Carmelo G., *Constructing the Ancient World: Architectural Techniques of the Greeks and Romans*, translated by Jay Hyams (Los Angeles: J. Paul Getty Museum, 2010), pp. 173–74.

"A Roman Bathhouse Still in Use after 2,000 Years," *BBC News Magazine*, October 13, 2013.

Walking Britain's Roman Roads, TV series, 2020, Channel 5 (UK).

Chapter 13. Chang'an (Trade)

Rothschild, Norman Harry, "Why Is It Necessary for Naked Savages to Drum and Dance? Early Tang Imperial Responses to a Sogdian Hibernal Festival," *Fudan Journal of the Humanities and Social Sciences* 8 (2015): 65–80.

UNESCO Silk Roads Programme, "Did You Know? The Cosmopolitan City of Chang'an at the Eastern End of the Silk Roads."

Chapter 14. Baghdad (Astronomy)

Akyol, Mustafa, "How We Lost Universalism" and "How We Lost the Sciences," in *Reopening Muslim Minds* (New York: St. Martin's Essentials, 2021), pp. 56–68, 86–104.

Sardar, Marika, "Astronomy and Astrology in the Medieval Islamic World," Metropolitan Museum of Art, 2011.

Scheiner, Jens J., and Damien Janos, eds., *The Place to Go: Contexts of Learning in Baghdad, 750–1000 C.E.* (Berlin: Gerlach Press, 2021).

Wiet, Gaston, *Baghdad: Metropolis of the Abbasid Caliphate*, translated by Seymour Feiler (Norman: University of Oklahoma Press, 1971).

Chapter 15. Kyoto (The Novel)

Columbia University, "What Is a *Waka*?," Asia for Educators, 2022.

Kiyoyuki, Higuchi, "Why Is There No Talk of Food or Bathing in the Tale of Genji?," in *Himitsu no Nihonshi* (*Secret History of Japan*), translated by Gregory Smits (Tokyo: Shōdensha, 1988), pp. 29–36.

Morley, Brendan Arkell, "Poetry and Diplomacy in Early Heian Japan: The Embassy of Wang Hyoryo˘m from Parhae to the Kōnin Court," *Journal of the American Oriental Society* 136, no. 2 (2016): 343–69.

Van Goethem, Ellen, "Why Leave the Nagaoka Capital?," *Nagaoka: Japan's Forgotten Capital* (Boston: Brill, 2008), pp. 237–52.

Chapter 16. Bologna (Universities)

Editors of *Encyclopedia Britannica*, "University of Bologna," *Encyclopedia Britannica*, July 18, 2019.

McSweeney, Thomas J., and Michéle K. Spike, "The Significance of the Corpus Juris Civilis: Matilda of Canossa and the Revival of Roman Law," William and Mary Law School, 2015, (*Faculty Publications*, 1736).

University of Bologna, "Nine Centuries of History." Also see the accompanying video, "Nove Secoli di Storia," YouTube, February 14, 2017.

Chapter 17. Hangzhou (Paper Currency)

Columbia University, "The Song Economic Revolution: From Copper Coins to Paper Notes," Asia for Educators.

Su Tung-Po, "On the Birth of a Son," Poetry Foundation.

Szczepanski, Kallie, "The Invention of Paper Money," ThoughtCo, 2019.

Weiner, Eric, "Genius Is Nothing New: Hangzhou," *The Geography of Genius: Lessons from the World's Most Creative Places* (New York: Simon & Schuster, 2016), pp. 65–96.

Chapter 18. Florence (Art)

Squires, Nick, "Renaissance Genius Raphael Revived Long-Lost Pigment Invented by Ancient Egyptians," *The Telegraph*, October 6, 2020.

Weiner, Eric, "Genius Is Expensive: Florence," *The Geography of Genius: Lessons from the World's Most Creative Places* (New York: Simon & Schuster, 2016), pp. 97–140.

Chapter 19. Dubrovnik (Public Health)

Alebić, Tamara, and Helena Marković, "Development of Health Care in Dubrovnik from 14th to 16th Century—Specific Features of Ragusan Medicine," *Collegium Antropologicum* 41, no. 4 (2017): 391–98.

Reed, Lawrence W., "Remembering the Ragusan Republic," Foundation for Economic Education, April 10, 2019.

Tomić, Zlata Blažina, and Vesna Blažina, *Expelling the Plague: The Health Office and the Implementation of Quarantine in Dubrovnik, 1377–1533* (Montreal: McGill-Queen's University Press, 2015).

Vuković, Kristin, "Dubrovnik: The Medieval City Designed around Quarantine," BBC, April 22, 2020.

Chapter 20. Benin City (Security)

Century Project, "Remarkable Historical Figures of Ancient Benin Kingdom," Google Arts & Culture website.

Hansberry, William Leo, "The Material Culture of Ancient Nigeria," *Journal of Negro History* 6, no. 3 (1921): 261–95.

Pearce, Fred, "The African Queen," *New Scientist*, September 11, 1999.

Chapter 21. Mainz (Printing Press)

Hammond, Alexander C. R., "Johannes Gutenberg," *Heroes of Progress* (Washington: Cato Institute, forthcoming).

Madhvi, Ramani, "How a German City Changed How We Read," BBC, May 8, 2018.

Chapter 22. Seville (Navigation)

Bergreen, Laurence, *Over the Edge of the World: Magellan's Terrifying Circumnavigation of the Globe* (New York: William Morrow, 2003).

Cavendish, Richard, "The Casa de Contratación Established in Seville," *History Today*, January 2003.

Pigafetta, Antonio, *The First Voyage Round the World*, translated by Henry Edward Stanley (London: Hakluyt Society, 1874).

Sazatornil, Blanca, anvd Alicia Suárez, "Elcano's Return (*El Regreso de Elcano*)," Naval Museum of Madrid, Google Arts & Culture website.

Chapter 23. Amsterdam (Openness)

"Amsterdam: Capital of the Golden Age," Official Guide for Visiting the Netherlands.

Braudel, Fernand, *The Wheels of Commerce: Civilization & Capitalism, 15th–18th Century*, (New York: Harper & Row, 1982).

McCloskey, Deirdre, *Bourgeois Dignity: Why Economics Can't Explain the Modern World* (Chicago: University of Chicago Press, 2011).

Chapter 24. Agra (Architecture)

"Architecture," Official website of the Taj Mahal.

Koch, Ebba, *The Complete Taj Mahal* (London: Thames & Hudson, 2006).

Stanberg, Susan, "A Pilgrimage to the Taj Mahal: A 'Poem in Stone,'" NPR, December 30, 2004.

Chapter 25. Cambridge (Physics)

Bailey, Simon, "The Hanging of the Clerks in 1209," BBC, December 18, 2009.

"Isaac Newton's Apple Tree," U.K. National Trust Woolsthorpe Manor website.

Zutshi, Patrick, "The Dispersal of Scholars from Oxford and the Beginnings of a University at Cambridge: A Study of the Sources," *English Historical Review* 127, no. 528 (2012): 1041–62.

Chapter 26. Paris (Enlightenment)

Follett, Chelsea, "Alcohol and Caffeine Created Civilization," *USA Today*, February 28, 2017.

Lilti, Antoine, "The Kingdom of Politesse: Salons and the Republic of Letters in Eighteenth-Century Paris," *Republic of Letters: Journal for the Study of Knowledge, Politics, and the Arts* 1, no. 1 (2009).

"Marie-Thérèse Rodet Geoffrin," *Encyclopedia Britannica*.

Pinker, Steven, *Enlightenment Now: The Case for Reason, Science, Humanism, and Progress* (New York: Viking, 2018).

Chapter 27. Edinburgh (Social Science)

Buchan, James, *Crowded with Genius: The Scottish Enlightenment; Edinburgh's Moment of the Mind* (New York: HarperCollins, 2003).

Edinburgh World Heritage, "City of Genius: The Scottish Enlightenment," classroom resources.

Thornton, Robert D., "The University of Edinburgh and the Scottish Enlightenment," *Texas Studies in Literature and Language* 10, no. 3 (1968): 415–22.

Weiner, Eric, "Genius Is Practical: Edinburgh," *The Geography of Genius: Lessons from the World's Most Creative Places* (New York: Simon & Schuster, 2016), pp. 141–84.

Chapter 28. Philadelphia (Liberal Democracy)

Declaration of Independence (1776).

Nash, Gary B., "Philadelphia: An Imperfect but Undeniable 'Cradle of Liberty,'" PBS, June 18, 2011.

Chapter 29. Vienna (Music)

Andrews, Evan, "What Is the Oldest Known Piece of Music?," History Channel, September 1, 2018.

Powell, Jim, "Ludwig van Beethoven's Joyous Affirmation of Human Freedom," Foundation for Economic Education, December 1, 1995.

Weiner, Eric, "Genius Is Contagious: Vienna on the Couch," *The Geography of Genius: Lessons from the World's Most Creative Places* (New York: Simon & Schuster, 2016), pp. 217–50.

Chapter 30. Manchester (Industrialization)

Follett, Chelsea, "Scrooge and the Reality of the Victorian Home," *American Spectator*, December 12, 2018.

"Manchester—The First Industrial City," London Science Museum website.

Nevell, Michael, "Dark Satanic Mills? The Archaeology of the World's First Industrial City," *Current Archaeology*, May 25, 2010.

Tupy, Marian L., "Market Capitalism Has Achieved What Karl Marx Always Wanted," *CapX*, July 13, 2018.

Chapter 31. London (Emancipation)

Hammond, Alexander C. R., "William Wilberforce," *Heroes of Progress* (Washington: Cato Institute, forthcoming).

"The Story of Africa: The End of Slavery," BBC World Service.

Chapter 32. Wellington (Suffrage)

Hammond, Alexander C. R., "Kate Sheppard," *Heroes of Progress* (Washington: Cato Institute, forthcoming).

Pickles, Katie, "Why New Zealand Was the First Country Where Women Won the Right to Vote," The Conversation, September 18, 2018.

Chapter 33. Chicago (Railroads)

Klein, Christopher, "When a Horse Raced against a Locomotive during the Industrial Revolution," History Channel, February 26, 2019.

Koziarz, Jay, "Transportation That Built Chicago: The Importance of the Railroads," Curbed Chicago, September 21, 2017.

Solomon, Brian, John Gruber, Chris Guss, and Michael Blaszak, *Chicago: America's Railroad Capital: The Illustrated History, 1836 to Today* (St. Paul, MN: Voyageur Press: 2014).

Chapter 34. Los Angeles (Cinema)

Lewis, Dan, "Thomas Edison Drove the Film Industry to California," *Mental Floss*, September 29, 2022.

Montoya, Yvonne, "The 'Mother of Hollywood' Thought the City Was Going to Be a Christian Utopia Free from Alcohol, Gambling and Prostitution," LAist, August 18, 2020.

Chapter 35. New York (Finance)

Knight, Thaya Brook, "Wall Street Offers Very Real Benefits," *USA Today*, May 26, 2015.

"New York after WWII," PBS, September 8, 2003.

Chapter 36. Hong Kong (Noninterventionism)

Monnery, Neil, *Architect of Prosperity: Sir John Cowperthwaite and the Making of Hong Kong* (London: London Publishing Partnership, 2017).

Tupy, Marian L., "Hong Kong and the Power of Economic Freedom," HumanProgress.org, March 7, 2016.

———, "Is This Goodbye for Hong Kong?," HumanProgress.org, June 9, 2020.

Chapter 37. Houston (Spaceflight)

Hsu, Tiffany, "The Apollo 11 Mission Was Also a Global Media Sensation," *New York Times*, July 15, 2019.

Kennedy, John F., address at Rice University on the nation's space effort, September 12, 1962.

Muir-Harmony, Teasel, "How Apollo 8 Delivered Christmas Eve Peace and Understanding to the World," *Smithsonian Magazine*, December 11, 2020.

Sparrow, Giles, "Apollo 11 Mission Control: The People behind the Moon Landing," BBC Sky, July 10, 2019.

Chapter 38. Berlin (Fall of Communism)

Ariely, Dan, Ximena Garcia-Rada, Katrin Gödker, Lars Hornuf, and Heather Mann, "The Impact of Two Different Economic Systems on Dishonesty," *European Journal of Political Economy* 59 (2019): 179–95.

"Berlin Wall," History Channel, March 21, 2021.

Gramlich, John, "East Germany Has Narrowed Economic Gap with West Germany since Fall of Communism, but Still Lags," Pew Research Center, November 6, 2019.

Kennedy, John F., remarks at the Rudolph-Wilde Platz, Berlin, June 26, 1963.

Chapter 39. Tokyo (Technology)

Beckley Michael, Yusaku Horiuchi, and Jennifer M. Miller, "America's Role in the Making of Japan's Economic Miracle," *Journal of East Asian Studies* 18, no. 1 (2018): 1–21.

Reed, Lawrence W., "What Caused Japan's Post-War Economic Miracle?," Foundation for Economic Education, August 26, 2022.

Chapter 40. San Francisco (Digital Revolution)

Newman, Katelyn, "San Francisco Is Home to the Highest Density of Billionaires," *U.S. News & World Report*, May 10, 2019.

Protin, Corey, Matthew Stuart, and Matt Weinberger, "Animated Timeline Shows How Silicon Valley Became a $2.8 Trillion Neighborhood," *Business Insider*, December 18, 2020.

Weiner, Eric, "Genius Is Weak: Silicon Valley," in *The Geography of Genius: Lessons from the World's Most Creative Places* (New York: Simon & Schuster, 2016), pp. 287–320.

Discussion Questions for Book Clubs & Classrooms

1. Why do you think progress tends to emerge from cities?

2. What do the cities featured in this book share in common and how do they differ?

3. The author cites relative peace, freedom, and large populations of people as important factors behind progress. What are some examples of cities where those things contributed to progress? And what are some exceptions?

4. What role has the freedom to discuss a diversity of ideas played in promoting human progress? How did intellectual debates drive progress in the cities profiled in the book? *(Such as Athens and Edinburgh.)*

5. What role has intercultural exchange played in promoting human progress? How did such exchange enrich the cities in the book? *(Such as Chang'an and Amsterdam.)*

6. What role has competition played in promoting human progress? How did competition bring about advances in the cities in the book? *(Examples might include the courtly literary competition in Kyoto or the international competition discussed in the Houston chapter.)*

7. What role have financial incentives played in promoting human progress? How did profits motivate progress in the cities in the book? *(Examples might include the role of financial patronage for Florence's artists and Vienna's musicians or how the profit motive brought about new companies in Tokyo and San Francisco.)*

8. What other factors do you think have been important in helping bring about progress historically?

9. The chapters discuss many different kinds of progress, including technological advancements, artistic achievements, and new institutions and policies. Are these different aspects of progress connected? If so, how?

10. Which cities built on the achievements in prior chapters? And which developments were not soon followed up on or exported as we might have hoped?

11. What do you think were the *most* significant innovations in the history of human progress? Which cities changed the world the most?

12. If you could visit one of the centers of progress during its heyday, which one would you visit and why? What would you bring back as a souvenir?

13. Many of the earlier centers of progress may have been advanced for the time period, but the author frequently notes that a modern person would be horrified by several aspects of the past. What are some ways in which modern life is superior to ancient times?

14. How, if at all, has this book changed your perspective?

15. If you could include an additional chapter, what city would you add? For what achievement?

Index

industry *(continued)*
  Hangzhou, 146, 148–49
  Hong Kong, 310
  Mainz, 182
  Manchester, 261, 264, 266–67
  New York City, 302
  printing, 181–88, 246, 248
  Tokyo, 338
Indus valley people, 20, 39, 41, 43, 100
information. *See* libraries
innovation, in Tokyo, 341
An Inquiry into the Nature and Causes of
  the Wealth of Nations (Smith), 241
instability
  Baghdad, 118
  Berlin, 329
  Chang'an, 113–14
  Edinburgh, 236
  Hong Kong, 311–12
  Kyoto, 125
  Mainz, 184, 187
intellectual discourse, in Paris, 229
intellectual men's clubs, in Edinburgh, 238
intercultural synthesis, in Agra, 213
in vitro fertilization, 226
Iraq
  Baghdad, 20–21, 117–24
  Ur, 65–71
  Uruk, 31–37, 66
Irnerius, 139–40
Iron Age, 328
irrigation system, in Jericho, 11
Isa, Ustad, 215
Isenburg, Diether von, 186
Isla Cerritos (Chichén Itzá), 77
Islam
  Agra, 211, 213
  Baghdad, 20, 117
  Chang'an, 113
  Seville, 193
Islamic Golden Age, 123
Italian Renaissance (1330–1550), 151,
  155, 157
Italy
  Bologna, 135–42
  Florence, 151–59
  Rome, 99–107

It Happened One Night (film), 298
I'timād-ud-Daulah (Baby Taj), 212

Jacobite rebellions, 236
Jahangir, Emperor, 211–12, 214
Jainism, 211
Jamaica, 273
Japan
  Kyoto, 125–34, 336
  Tokyo, 335–42
The Jazz Singer (film), 297
Jefferson, Thomas, 238–39, 240, 246, 350
Jericho, 7–13
jiaozi. *See* paper currency
JIT. *See* just-in-time (JIT) manufacturing,
  in Tokyo
Jobs, Steve, 348
John, King of England, 221
John VIII, Pope, 271
Johnson, Lyndon B., 321
Joseph II, King of Austria, 255
Judaism, 21, 92, 113–14, 153, 168,
  203–4
Julius II, Pope, 156
The Jungle (Sinclair), 290
Justinian I, 166
Justinian the Great, 138
just-in-time (JIT) manufacturing, in
  Tokyo, 341

Kahun gynecological papyrus, 61
Kamo River (Kyoto), 126
kangaroos, at Budj Bim, 26
Kanmu, Emperor, 127–28
Kashani, Kalim, 217
Kennedy, John F., 321–22, 332
Khan, Genghis, 123
Khan, Hulagu, 123
Khrushchev, Nikita, 330
Ki family (Kyoto), 129
Kingdom of Francia, 229
Kingdom of Prussia, 328
King's College Chapel (Cambridge), 220
Kiyomaro, Wake no, 127
Kobe, Japan, 337
Kokcha River (Afghanistan), 68
Komachi, Ono no, 131

yam daisy. *See* murnong
Yamuna River, India, 209, 210, 216
Yangzhou massacre (760), 114
Yoshida, Shigeru, 339
Yuan Zhen, 112–13
Yucatán Peninsula, Mexico, 74–77

Zachary, Pope, 270–71
zaibatsu, 339
Zanzibar, 276
Zersetzung, 331
zoning regulations, in Houston, 318
Zoroastrianism, 113, 211

About the Author

Chelsea Follett is a policy analyst at the Cato Institute's Center for Global Liberty and Prosperity and the managing editor of HumanProgress.org, a project of the Cato Institute that seeks to educate the public on the global improvements in well-being by providing free empirical data on long-term developments. Her writing has been published in the *Wall Street Journal, USA Today, Newsweek, Forbes, The Hill, Business Insider, National Review, Washington Examiner, Reason,* the *Richmond Times Dispatch,* and the *Virginian-Pilot,* among other outlets. She was named to *Forbes's* 30 under 30 list for 2018 in the category of law and policy. Follett earned a BA magna cum laude in government and English from the College of William & Mary, as well as an MA in foreign affairs from the University of Virginia, where she focused on international relations and political theory. She lives in Virginia with her husband and two children.

About the Cato Institute & HumanProgress.org

Founded in 1977, the Cato Institute is a public policy research foundation dedicated to broadening the parameters of policy debate to allow consideration of more options that are consistent with the principles of limited government, individual liberty, and peace. The Institute is named for *Cato's Letters*, libertarian pamphlets that were widely read in the American colonies in the early 18th century and played a major role in laying the philosophical foundation for the American Revolution.

The Cato Institute undertakes an extensive publications program on the complete spectrum of policy issues. Books, monographs, and shorter studies are commissioned to examine the federal budget, Social Security, regulation, military spending, international trade, and myriad other issues. Major policy conferences are held throughout the year.

HumanProgress.org, a project of the Cato Institute, was founded in 2013 to rectify the widely held misperceptions about the state of humanity by gathering empirical data from reliable sources that look at worldwide long-term trends. By compiling and presenting these comprehensive data in an accessible way, the editors aim to provide a useful resource for students, scholars, journalists, policymakers, and the general public.

To maintain its independence, the Cato Institute accepts no government funding. Contributions are received from foundations, corporations, and individuals, and other revenue is generated from the sale of publications. The Institute is a nonprofit, tax-exempt, educational foundation under Section 501(c)(3) of the Internal Revenue Code.